COMPUTER PROBLEMS FOR CLASSICAL DYNAMICS

AN INTEGRATED APPROACH

CHARLES LEMING
HENDERSON STATE UNIVERSITY

HARCOURT BRACE JOVANOVICH, PUBLISHERS
and its subsidiary, Academic Press

SAN DIEGO   NEW YORK   CHICAGO   AUSTIN   WASHINGTON, D.C.
LONDON   SYDNEY   TOKYO   TORONTO

Copyright © 1988 by Harcourt Brace Jovanovich, Inc.

All rights reserved. No part of this publication may
be reproduced or transmitted in any form or by any
means, electronic or mechanical, including photocopy,
recording, or any information storage and retrieval system,
without permission in writing from the publisher.

Requests for permission to make copies of any part of the
work should be mailed to: Permissions, Harcourt Brace
Jovanovich, Publishers, Orlando, Florida 32887.

ISBN: 0-15-507635-3

Printed in the United States of America

Preface

The value of mathematics as an aid to describing and understanding nature was first recognized several thousand years ago. Since that time, all physicists have shared a common plight: the difficulty of communicating and interpreting mathematical ideas. Although the earliest computers provided new ways of doing things, they contributed little to the solution of this problem. It took the development and widespread distribution of graphics computers to provide practical and interactive visual displays of the results of calculations -- changing forever the methods of using mathematics in physics.

The purpose of this text is to enhance the study of classical mechanics at the intermediate undergraduate level by developing the skills and concepts necessary to successfully apply computers to problem solving and communication. To accomplish this goal, students use the programs of this text as a starting point for creating their own interactive graphical displays through a synthesis of numerical methods and computer graphics. By stressing involvement and activity in a laboratory setting, students can both acquire new skills and gain an improved understanding of classical mechanics.

I was absolutely intrigued by computer graphics from my first encounter with a computer equipped with a high resolution color display. When a terminal was installed near my office, I regularly found excuses to experiment with three-dimensional drawings and representations of mathematical functions as a break from my real work. After a few practical uses were found to justify my play, it gradually dawned on me that this skill could be a valuable tool for students both in the study of physics and in their subsequent professional careers.

Many of the programs in this text were first developed during class periods and then tested by students in the physics classes that I teach at Henderson State University. The development of this text would have been impossible without the help, patience, and inspiration of my students. I am also especially grateful to my colleagues Dr. Donald A. Avery and Dr. Bryan D. Palmer for their encouragement and suggestions.

In addition, I wish to acknowledge the help of my wife Paula and my daughter Sarah. Their assistance with language and wording was a tremendous help and their moral support was even more valuable.

Finally, I wish to acknowledge the assistance of Dr. Stephen T. Thornton of the University of Virginia and Mr. Jeff Holtmeier of Harcourt Brace Jovanovich. Without their initial interest, I might never have conceived of the possibility of a textbook on this subject.

Charles W. Leming

CONTENTS

Chapter 1. Introduction

1.1 Problems and Solutions 1
1.2 To the Instructor 3
1.3 To the Student 5
1.4 References 6

Chapter 2. Rotation of Coordinates

2.1 Introduction 7
2.2 Rotation of Coordinates in Two Dimensions 8
2.3 Rotation of Coordinates in Three Dimensions 12
   Exercises 22
   Problems 22

Chapter 3. Euler's Method and Associated Numerical Techniques

3.1 Introduction 24
3.2 Euler's Method 25
3.3 Half-Step Approximation 27
3.4 Last Point Approximation 28

Chapter 4. Motion in One Dimension

4.1 Introduction 30
4.2 Euler's Method 31
4.3 Computer Program 31
4.4 Graphics Program 34
4.5 Scaling the Figure 34
4.6 Rockets 37
4.7 Two-Stage Rockets 39
4.8 Simple Pendulum 42
   Exercises 46
   Problems 47

Chapter 5. Motion in Two Dimensions

5.1 Introduction 49
5.2 Harmonic Oscillations in Two Dimensions 49
5.3 Foucault Pendulum 52
5.4 Central Force Motion (Two-Body Motion in Polar Coordinates) 53
5.5 Orbital Boost 57
5.6 Central Force Motion (Cartesian Coordinates) 59
5.7 Three-Body Motion (Two Dimensions) 62
5.8 Classical Hall Effect 62
   Exercises 65
   Problems 66

CONTENTS

Chapter 6. Motion in Three Dimensions

6.1 Introduction	68
6.2 Motion of a Projectile in a Rotating Coordinate System	69
6.3 Motion of a Symmetrical Top	76
Exercises	83
Problems	83

Appendix A Computer Graphics

A.1 Introduction	85
A.2 Screen Concepts	86
A.3 Graphics Operations	86
A.4 Graphing a Mathematical Function (Cartesian Coordinates)	88
A.5 Scaling and Positioning the Figure	89
A.6 Graphing a Mathematical Function (Polar Coordinates)	90
A.7 More BASIC Language	91
A.8 Program Outline	92
A.9 Graphics Style Manual	93
Exercises	93

Appendix B Summary of Graphics Operations

B.1 Introduction	97
B.2 Beagle Graphics	98
B.3 GW-BASIC	99
B.4 DEC ReGIS	100

Appendix C The Fourth Order Runge-Kutta Method

C.1 Introduction	103
C.2 Applying the Fourth Order Runge-Kutta Method	103

Appendix D Program Listings for the IBM-pc

Computer Problems for Classical Dynamics: An Integrated Approach

CHAPTER 1

Introduction

1.1 Problems and Solutions

Traditionally, most students begin the study of classical mechanics at a period in their careers somewhere between "introductory" and "advanced" physics. (See *Classical Dynamics* by Jerry Marion, Preface to the First Edition.) Therefore, the level of the material in this text is between "introductory" and "advanced," allowing students at the intermediate stage to gain experience using microcomputers as problem solving tools. The materials are designed to accompany the text *Classical Dynamics* (third edition) by Marion and Thornton, but are suitable for use with any standard undergraduate dynamics text.

Because a course in classical mechanics often forms the basis for further study in several areas, the general utility of computers in theoretical physics is emphasized. Complex sets of differential equations often result from application of the formal procedures of classical mechanics. These equations can be evaluated using a computer to perform the repetitive calculations required to apply numerical techniques. Students planning careers in engineering or in secondary education will find these techniques particularly useful.

Using these procedures, mathematical investments yield much greater returns. Even students having minimal mathematical training can readily solve and understand some of the most elegant and interesting problems of classical mechanics. Graphical display of the solutions permits a clear view of physical principles with a notable reduction in mathematical fog.

The development of this text required a long series of decisions and compromises. Some individuals have strong opinions about matters such as the type of computer, the choice of programming language and the order of numerical solutions; however, only a few such choices are of fundamental importance. Most of the decisions are arbitrary and can be changed to suit the local situation. It is hoped that this series of decisions and compromises has resulted in a text which is interesting, useful and flexible.

The original versions of the programs contained in this text were developed using the DEC ReGIS graphics language and operated on DEC and IBM minicomputers. The programs were later rewritten for use with the Apple IIe computer and an inexpensive graphics package (Beagle Graphics) because of the wide availability and low cost of this system. Appendix B and Appendix D contain instructions for converting the programs for use with DEC ReGIS or with GW-BASIC used with IBM-pc "compatible" computers.

Programs in this text make extensive use of graphics. In fact, the ability to create a graphical representation of calculations is the greatest advantage of the use of computers in the study of classical mechanics. The near-universal availability of graphics capability with modern computers has supplied the missing ingredient without which computers were formerly of only limited utility for problem solving and communication in physics. With the use of graphics, numerical results can be translated into intuitive understanding. For example, programs which allow students to solve problems in three dimensions and to display the results in any desired orientation are developed in Chapter 2. The programs are then applied to display the motion of three- dimensional systems in Chapter 6.

A survey of numerical methods reveals that many techniques exist for solving problems of the type encountered in classical mechanics. Some of the methods are useful for specialized types of problems while others are general in their application. Euler's Method (following from the familiar definition of the derivative), the half-step approximation (introducing the idea of successive improvements in numerical schemes), and the last point approximation (assuring limits on the error in many problems) provide an adequate introduction to numerical methods. The last point approximation is particularly important as it allows reasonable accuracy even with simple numerical methods. Because sufficient accuracy is obtainable with these methods, a discussion of higher order approximations is not included in the body of the text.

If higher accuracy is desired, Appendix C contains a "recipe" for applying the fourth order Runge-Kutta method. Because they are simple and easy to apply, it is best for students to begin by applying the methods described in the body of the text. Using these methods, students also acquire experience with the errors and limitations of numerical methods. Higher order methods can produce such accurate results that students might not gain an appreciation of these problems.

Other techniques such as methods of solving boundary value problems and Monte-Carlo methods were also omitted because of the limited goals of this text. Students can later be introduced to these ideas in a mathematical physics course at the advanced undergraduate level.

Another important decision involved the choice of units for physical quantities. It is not unusual to begin theoretical calculations by defining reduced variables. (Every physicist is familiar with the statement: "Let $e = h = c = 1$.") Used sparingly, reduced variables can sometimes reduce confusion by allowing advanced students to understand physical principles without depending on a specific system of units.

Because it is both easy and useful to scale the computer screen in any system of units (see Appendix A), the use of reduced variables was avoided. SI units and cgs units are used throughout these materials. For students at the intermediate level, the results of calculations are more realistic and satisfying when figures are scaled using a standard system of units.

After these decisions were made, the fun could begin. However, even the fun of solving dynamics problems had to be restricted. The problem of selecting examples from the endless number of possibilities was solved by following the general outline of a traditional and well accepted text.

1.2 To the Instructor

It is clear from the preceding section that this text is not meant to be the last word in computational physics, nor is it intended to provide a complete course in classical mechanics. This text is intended as a supplement to enhance a traditional course in classical mechanics. The text is, however, general enough to serve as reference for applying these methods in a context other than the study of classical mechanics.

The topics covered can be easily integrated into a traditional course (usually 3 credit hours per semester for two semesters). If all of the supplemental materials are used, a few traditional topics might be deleted. The individual instructor may decide how to structure the course under these circumstances. Instead of integrating the material covered in this text into a traditional course, it is also possible to offer a computer course with classical mechanics as a corequisite or prerequisite (usually 1 credit hour per semester for two semesters). An instructor might wish to implement this option before attempting the more difficult problem of integrating computer methods into the framework of an already existing classical mechanics course.

A list of the topics covered in each chapter is included in the table of contents. References in the body of the text key each section to the appropriate section of the text *Classical Dynamics* by Marion and Thornton. The topics in this text can then be covered in whatever order they arise in the context of a traditional course. It will, however, be necessary to devote some time to a discussion of elementary numerical methods (Chapter 3). These techniques are not ordinarily covered in classical mechanics courses and students may not be familiar with them. A brief discussion of the BASIC programming language could be included at the beginning of the course. As the course proceeds, even the uninitiated can easily acquire the small amount of required programming knowledge without further formal instruction.

Computer graphics may present more of a problem to students. It is suggested that the exercises in Appendix A be used to develop a base of common experience before proceeding with the material in the body of the text. Any computer capable of graphics can be utilized for these exercises. However, a computer with low graphics resolution or which is incapable of mixed graphics and text may prove to be unsatisfactory for generating the more complex and detailed figures. Because graphics instructions are not standardized, the syntax of the graphics statements used in the example programs must be modified to suit each specific type of computer. Appendix B contains information on how to convert the programs for use with other types of computers.

Exercises are included at the ends of chapters in order to allow students to test their understanding. Students can quickly complete most of the exercises by altering a few lines of the programs included in the text. In addition to the exercises, a limited selection of problems is included at the end of each chapter. Problems require the application of newly learned techniques to unfamiliar situations. Solving the problems often requires writing original programs rather than modifying existing programs as in the exercises.

Because many formerly intractable problems become simple when computer methods are employed, students can explore the dynamics of a widely expanded assortment of systems. So many possibilities exist that students can often complete individual projects instead of working standard problem sets assigned to the entire class. Communication techniques can be developed and skills improved by having students present the results of projects (complete with drawings and computer programs) at informal seminars or in written papers, rather than merely handing in homework to be graded and returned.

Experience indicates that students learn best when encouraged to experiment in a directed "laboratory" setting. As a result, the decision was made to provide complete versions of all programs in the body of the text. All programs follow a standard outline. Rather than structure programs for greatest efficiency, programs relate as clearly as possible to the equations of motion of the physical system being studied. Any program in this text could be modified for more efficiency in computing and for increased speed and accuracy. In addition, any of the programs can be "crashed" by using extreme values for variables.

The main danger inherent in the decision to include complete programs is that the programs might simply be used for classroom "demonstrations" rather than to provide hands-on "laboratory" experiences for students.

Many famous physicists have expressed the satisfaction and thrill of discovering previously hidden physical principles for themselves. (For an enjoyable and humorous modern example see *Surely You're Joking, Mr. Feynman!* by Richard Feynman.) Today's physics students can experience the feeling of discovery as they explore the properties of unique systems which they "invent" and study. Theoretical physics becomes similar to experimental physics as students perform numerical "experiments" in order to answer questions about the systems which they are studying. Understanding is promoted by making changes in systems and testing the results against intuitive predictions.

1.3 To the Student

The effective use of the programs in this text requires three broad categories of activities: entering and de-bugging programs, experimentation, and addition of "bells and whistles."

The first computer bug was reportedly a moth found jamming a mechanical relay in the Mark II computer at Harvard on September 9, 1947. Since that time, many long hours have been spent de-bugging programs. Students who use this text will spend some time pursuing this activity. The search for bugs is not the trivial waste of time that it may first appear. While encoding and de-bugging programs, students examine the inner workings and foundations of the programs. Because of this, students should enter programs for themselves instead of relying on copies from other sources.

After a program is operating correctly, students should experiment and test the limits of the system that they are studying. This is the heart of the methods of this text. Unlike conventional homework sets which are an end product to be handed in and graded, computer programs are just a beginning.

The computer programs in this text are written in a "bare bones" fashion using no specialized programming statements. Students should feel free to add their own personal touches such as color graphics, sound, and labels. As with the other activities noted above, the addition of these seemingly frivolous "bells and whistles" creates involvement while promoting understanding and communication.

The computer has changed the study of classical mechanics just as the laser has changed the study of optics. A very few years ago, both subjects were considered to be outside the mainstream of modern research. In both cases, technological advance brought renewed interest and possibilities for fundamental discoveries.

In mechanics, theories resulting from numerical analysis are yielding new understanding of chaos. New computational algorithms are being developed which efficiently solve the equations resulting from the analysis of complex systems. As more complex equations of motion are subjected to numerical analysis, new schemes are being developed in the hope of simplifying the formulation of equations describing complex systems.

Both students and instructors will find *American Journal of Physics* to be a valuable source of new ideas. With the renewed interest in classical mechanics, articles which develop the equations of motion for complex mechanical systems appear regularly in this journal (often with thoroughly developed analytical solutions which can be compared to numerical results). Students should be read such articles as a source of recent ideas in the field of classical mechanics.

1.4 References

A partial list of recent texts of special interest is given below.

W. Newman and R. Sproll, *Principles of Interactive Computer Graphics* (McGraw-Hill, New York, 1979)

J. Foley and A. van Dam, *Fundamentals of Interactive Computer Graphics* (Addison-Wesley, Reading, MA, 1981)

J. Danby, *Computing Applications to Differential Equations* (Reston, Reston, VA, 1985)

S. Koonin, *Computational Physics* (Benjamin/Cummings, Menlo Park, CA, 1986)

J. Marion and S. Thornton, *Classical Dynamics of Particles and Systems*, Third Edition (Harcourt Brace Jovanovich, San Diego, CA, 1988)

K. Symon, *Classical Mechanics* (Addison-Wesley, Reading, MA, 1974)

D. Davis, *Classical Mechanics* (Academic Press, San Diego, CA, 1986)

G. Fowles, *Analytical Mechanics* (Holt, Rinehart and Winston, New York, 1985)

CHAPTER 2

ROTATION OF COORDINATES

2.1 Introduction

Coordinate systems and coordinate transformations are a major preoccupation of physicists. Finding the right coordinate system and specifying the best point of view is often the key to describing and understanding physical systems. For example, plane polar coordinates (r,θ) are often used to describe physical systems with circular symmetry. Three-dimensional systems with spherical symmetry are also represented using spherical coordinates (r,θ,φ). (See Marion and Thornton, Appendix F.)

Mathematical functions that are written in terms of plane polar coordinates are graphed on a computer screen by converting the coordinate locations of the points of the polar function to cartesian screen locations. (See Appendix A, section A.6.) This technique is applied in Program 2.1 to plot a graph of Eq. 2.1, the polar equation for a "three leaved rose."

$$r = r_0 \sin(3\theta) \qquad (2.1)$$

The program plots the function by calculating the values of r as the angle θ is varied from 0 to 3.14 radians in increments of 0.01 radians. (The angle θ is represented by the variable A in this program.) The cartesian coordinates (x, y) corresponding to the polar coordinates (r, θ) found in each successive calculation are calculated in line numbers 1030-1040 using the relations: $x = r\cos\theta$ and $y = r\sin\theta$. Applying the graphical techniques described in Appendix A, the screen coordinates (XS, YS) corresponding to this point are then determined in line numbers 1070-1080 using statements of the form of Eq. A.2 and Eq. A.3. Line number 1090 plots the point by displaying the screen pixel located at the screen position (XS, YS).

```
90   REM : PROGRAM (2.1)
100  REM
              ***** SET UP GRAPHICS CHARACTERISTICS *****

110  & HGR2 : & HCOLOR= 15: & B COLOR= 0: & PRINT : & MODE(1)
300  REM
                    *****  SET UP SCREEN DISPLAY *****

310  &  HPLOT 1,96 TO 550,96
320  &  HPLOT 280,1 TO 280,190
330  & GOTO 440,12:PRINT"R=R0*sin(3*A)"
500  REM
                  ***** SPECIFY INITIAL CONDITIONS *****

510 R0 = 90
1000 REM

                    *****  PLOT POLAR FUNCTION *****

1010 FOR A = 0 TO 3.14 STEP .01
1020 R = R0 *  SIN (3 * A)
1030 X = R *  COS (A)
1040 Y = R *  SIN (A)
1070 XS = 280 + 2 * X:REM GRAPHICS SCREEN HAS AN ASPECT RATIO OF 2
1080 YS = 96 - Y
1090   &  HPLOT XS,YS
1110   NEXT A
2000   END
```

2.2 Rotation of Coordinates in Two Dimensions

The figure formed by plotting Eq. 2.1 can be rotated on the sceen by applying the transformation equations for the rotation of coordinates. (See Marion and Thornton, Section 1.3.)

$$x_1 = x\cos(\theta_1) + y\sin(\theta_1) \tag{2.2}$$

$$y_1 = -x\sin(\theta_1) + y\cos(\theta_1) \tag{2.3}$$

The quantities x_1 and y_1 [X1 and Y1 in Program 2.2] are the values of the coordinates x and y relative to the new (rotated) coordinate

2.2 ROTATION OF COORDINATES IN TWO DIMENSIONS 9

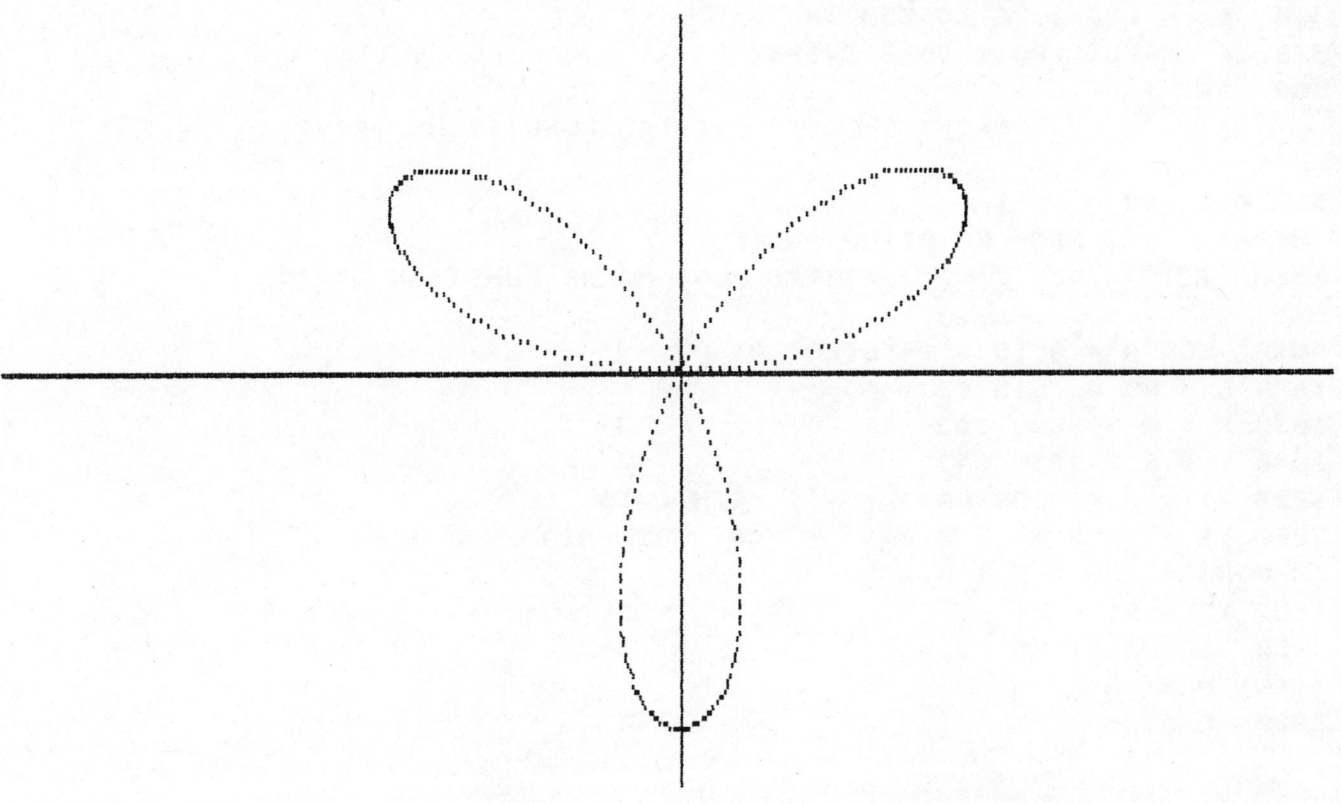

Figure 2.1 Graph of the figure defined by the polar function, Eq. 2.1.

system. Application of these transformation equations is equivalent to a rotation of the coordinate axes through an angle θ_1 [A1 in Program 2.2]. The procedure employed in Program 2.2 always creates the rotated coordinate axes in the same position on the screen independent of the value of θ_1. Thus a rotation of the coordinate axes through an angle θ_1 causes the graph of the function to appear on the screen as if it were rotated through an angle $-\theta_1$.

Line numbers 1030-1040 apply the transformation equations which calculate X1 and Y1 which are the coordinates of the points of the figure relative to the rotated coodinate system. The coordinates X1 and Y1 are then converted to screen coordinates and plotted as in Program 2.1.

```
90   REM : PROGRAM (2.2)
100  REM
                ***** SET UP GRAPHICS CHARACTERISTICS *****

110  & HGR2 : & HCOLOR= 15: & B COLOR= 0: &  PRINT : & MODE(1)
300  REM
                *****  SET UP SCREEN DISPLAY *****
```

```
310  &  HPLOT 1,96 TO 550,96
320  &  HPLOT 280,1 TO 280,190
500  REM
                    ***** SPECIFY INITIAL CONDITIONS *****

510 R0 = 90
520 A1 = .3: REM  ROTATION ANGLE
1000  REM                     *****  PLOT POLAR FUNCTION *****

1010  FOR A = 0 TO 3.14 STEP .01
1020  R = R0 *  SIN (3 * A)
1030  X = R *  COS (A)
1040  Y = R *  SIN (A)
1050  X1 = X *  COS (A1) + Y *  SIN (A1)
1060  Y1 =  - X *  SIN (A1) + Y *  COS (A1)
1070  XS = 280 + 2 * X1
1080  YS = 96 - Y1
1090  &  HPLOT XS,YS
1110  NEXT A
2000  END
```

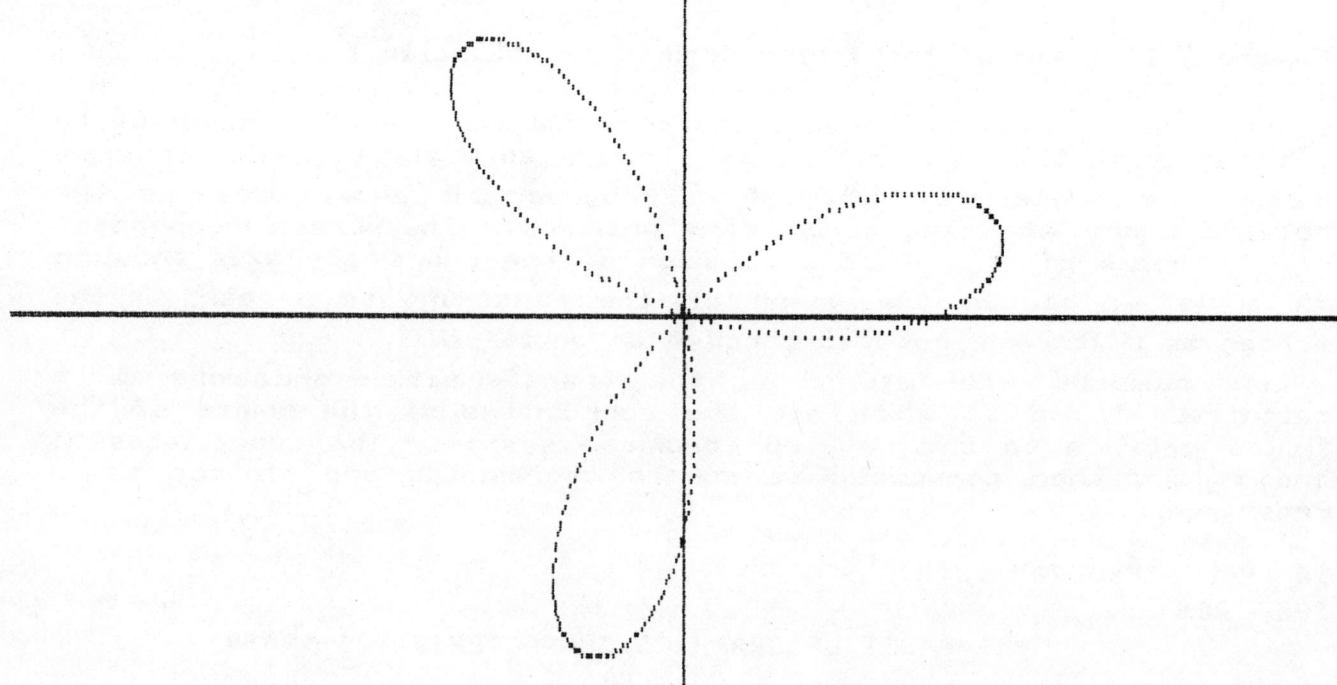

Figure 2.2 A graph of the figure defined by the polar function Eq. 2.1 is drawn relative to coordinate axes rotated by 0.3 radians. Note that the figure is rotated on the screen by an angle of -0.3 radians.

2.2 ROTATION OF COORDINATES IN TWO DIMENSIONS

Polar functions can also be written as sets of parametric equations. For example, Eq. 2.4 and Eq. 2.5 are parametric equations describing an ellipse whose axes lie along the coordinate axes.

$$x = A\cos(\theta) \tag{2.4}$$

$$y = B\sin(\theta) \tag{2.5}$$

The values of A and B determine the eccentricity of the ellipse. If A and B are equal, the figure described by the equations is a circle of radius A. Program 2.3 plots an ellipse based on Eq. 2.4 and Eq. 2.5.

```
90   REM   PROGRAM (2.3)
100  REM
                 ***** SET UP GRAPHICS CHARACTERISTICS *****

110  &  HGR2 : &  HCOLOR= 15: &  B COLOR= 0: &  PRINT : & MODE(1)
300  REM
                 *****  SET UP SCREEN DISPLAY *****

310  &  HPLOT 1,96 TO 550,96
320  &  HPLOT 280,1 TO 280,190
500  REM
                 ***** SPECIFY INITIAL CONDITIONS *****

510  A = 90
520  B = 30
1000 REM
                 *****  CALCULATE VALUES AND PLOT FUNCTION *****

1010  FOR TH = 0 TO 6.28 STEP .02
1020  X = A *  COS (TH)
1030  Y = B *  SIN (TH)
1040  XS = 280 + 2 * X
1050  YS = 96 - Y
1060  &  HPLOT XS,YS
1070  NEXT TH
2000  END
```

2 · ROTATION OF COORDINATES

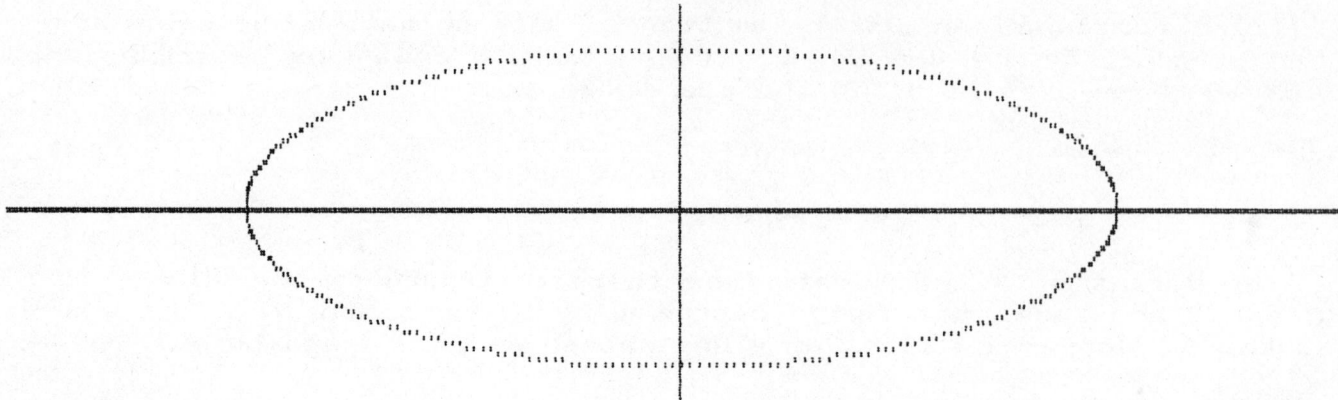

Figure 2.3 An ellipse based on Eq. 2.4 and Eq. 2.5 is drawn using Program 2.3. In this figure A=3B.

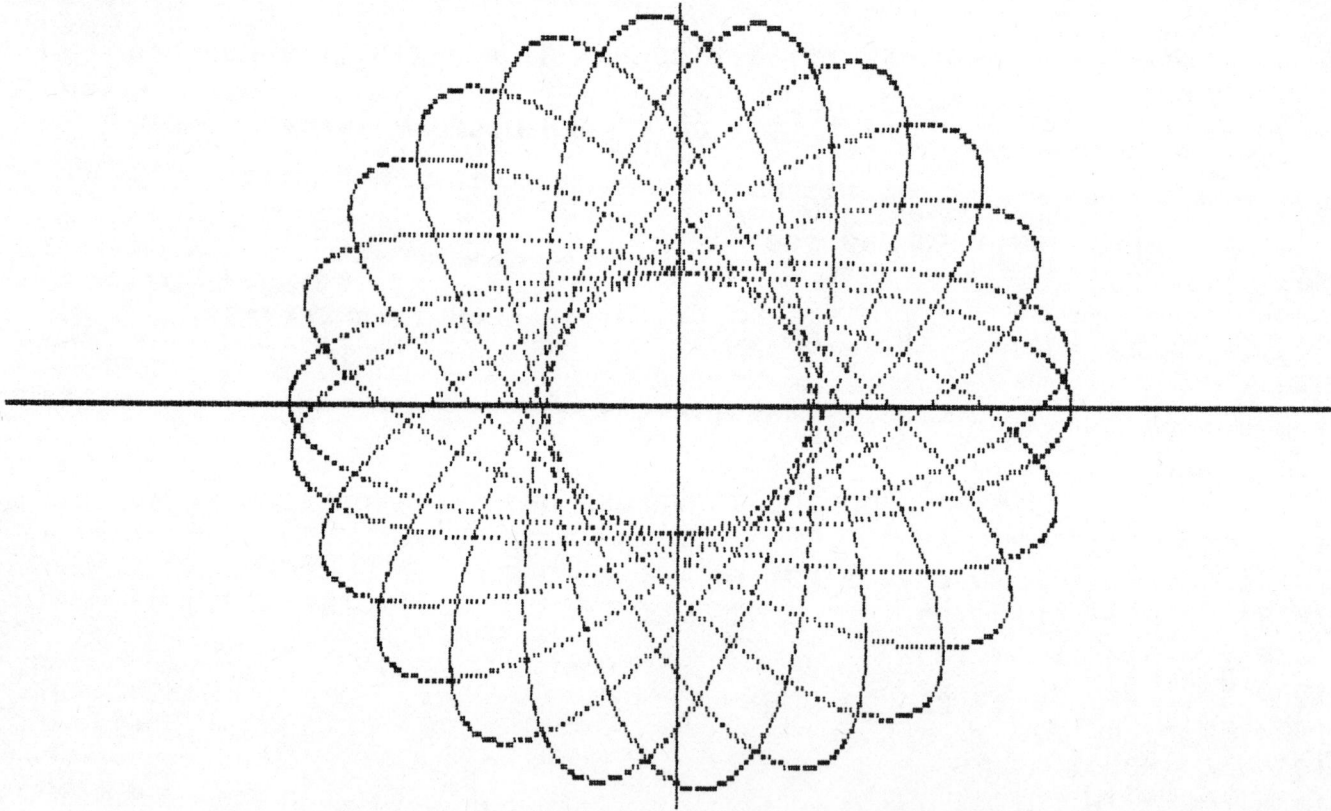

Figure 2.4 A series of ellipses is drawn with each succeeding ellipse rotated through an angle of 0.3 radians by applying the transformation equations for rotation of coordinates (Eq. 2.2 and Eq. 2.3).

2.3 Rotation of Coordinates in Three Dimensions

The techniques used to plot graphs of two-dimensional functions can be extended to three-dimensional functions. Figures described by three-dimensional functions can be plotted on the two-dimensional computer screen using a point by point projection of the figure onto the screen. The three-dimensional figure can then be displayed in any

2.3 ROTATION OF COORDINATES IN THREE DIMENSIONS

orientation on the computer screen by means of a technique similar to that developed in Program 2.2 for the rotation of two-dimensional figures.

Rotation of a three-dimensional figure is best described by successive, independent rotations in each of three planes. A set of three linear equations specifying the transformation properties of the coordinates of a point can then be written in terms of the direction cosines for each rotation. This set of equations is expressed in terms of a transformation matrix. (See Marion and Thornton, Section 1.3.) The combination of rotations is thus described by a single three-dimensional transformation matrix formed by an ordered multiplication of the transformation matrices describing the rotation in each plane. (See Marion and Thornton, Section 1.7.) Although many possible sets of three independent angles can be used to transform one coordinate system to another, the program below is based on the transformation matrix using Eulerian angles (φ, θ, ψ). (See Marion and Thornton, Section 10.7.)

In order to perform the transformation, the elements of the transformation matrix are first calculated using the specified values of φ, θ, and ψ. (Line numbers 110-130 specify these angles in radians and line numbers 200-280 determine the values of the elements of the transformation matrix.) The coordinates of the viewer are calculated in line numbers 300-320 using the variable D to specify the distance between the viewer and the screen. A translation of coordinates is then performed to move a point (x,y,z) into a reference system in which the viewer is at the origin (line numbers 2000-2020). This translation assures that the image will remain centered at the specified point on the screen after being rotated.

After the translation is performed, the location of the point in the original system (X,Y,Z) is converted to a location in the rotated coordinate system (X3,Y3,Z3) by applying the rotation matrix (line numbers 2030-2050). The new coordinate values (X3, Y3, Z3) are projected onto a two-dimensional plane (the computer screen) a distance D from the viewer (line numbers 2070-2080). Screen coordinates (XS, ZS) are calculated in this step so that the image is drawn to the correct scale and centered on the screen.

A real object can be defined in terms of a series of coordinate locations (x, y, z). The image resulting from the rotation and projection of these points onto the computer screen is a perspective drawing of the object defined by the series of points. The screen of the computer monitor can then be thought of as a window through which the drawing is viewed.

The Eulerian angles can be specified to provide an arbitrary view of the three-dimensional object. If the values of the Eulerian angles are all equal to zero, the y axis is horizontal and the z axis is vertical (the screen represents the y-z plane). In this orientation, the x axis must be imagined projecting outward from the screen. By selecting different values of φ, θ, and ψ (A1, A2, and A3), the axes are rotated by the appropriate amount and the origin of coordinates remains centered on the screen. For example, the screen becomes the x-y plane and the z axis projects from the screen if the values chosen for φ, θ, and ψ are: A1 = 1.57, A2 = 0, and A3 = 1.57 respectively.

2 · ROTATION OF COORDINATES

The DATA statements in Program 2.4 define a square, flat surface in the x-y plane. [The coordinates of the corners of this plane are (3, 3, 0), (3, -3, 0), (-3, -3, 0), and (-3, 3, 0).] Each side of the surface is thus six units in length. Each time the READ statement in line number 330 is executed, coordinates of the end points of a line in the grid on the x-y plane are found. The projection of these points onto the computer screen is performed using the rotation subroutine. A line is then drawn between the points (line number 500). By changing the values of the Eulerian angles (A1, A2, and A3) and by changing the viewing distance (D) the drawing will change as the apparent position of the viewer is altered relative to the x-y plane. The origin of coordinates of the rotated surface always remains at the center of the screen as the surface is rotated.

```
7   REM     PROGRAM (2.4)
10  REM     THIS PROGRAM USES EULER'S ANGLES FOR ROTATION
80  REM
                    ***** SET UP GRAPHICS CHARACTERISTICS *****


90  & HGR2 : &  HCOLOR= 15: & B COLOR= 0: &  PRINT : & MODE(1)
95  REM
                    ***** SET UP SCREEN DISPLAY *****
100 REM         ANGLE A1 IS PHI:REM ANGLE A2 IS THETA:REM ANGLE A3 IS
PSI:REM ALL ANGLES ARE EXPRESSED IN RADIANS
110 A1 = 0
120 A2 = 0
130 A3 = 0
135 &   GOTO 510,10: PRINT "A1=";A1
137 &   GOTO 510,18: PRINT "A2=";A2
139 &   GOTO 510,26: PRINT "A3=";A3
140 S1 =  SIN (A1):C1 =  COS (A1)
150 S2 =  SIN (A2):C2 =  COS (A2)
160 S3 =  SIN (A3):C3 =  COS (A3)
170 REM                                                   L1  L2  L3
180 REM     CALCULATE ELEMENTS OF TRANSFORMATION MATRIX   L4  L5  L6
190 REM                                                   L7  L8  L9
200 L1 = C3 * C1 - C2 * S1 * S3
210 L2 =  - S3 * C1 - C2 * S1 * C3
220 L3 = S2 * S1
230 L4 = C3 * S1 + C2 * C1 * S3
240 L5 =  - S3 * S1 + C2 * C1 * C3
250 L6 =  - S2 * C1
260 L7 = S3 * S2
270 L8 = C3 * S2
280 L9 = C2
```

2.3 ROTATION OF COORDINATES IN THREE DIMENSIONS

```
290 D = 20: REM            DISTANCE OF VIEWER FROM SCREEN (CM)
300 XV = D * L1
310 YV = D * L2
320 ZV = D * L3
330  READ A,X1,Y1,Z1,X,Y,Z
340  IF A = 16 GOTO 5000
350  GOSUB 2000
490 SX = XS:SZ = ZS:X = X1:Y = Y1:Z = Z1
495  GOSUB 2000
500  &  HPLOT XS,ZS TO SX,SZ
510  IF A = 4 THEN  &  GOTO XS,ZS: PRINT "X"
520  IF A = 11 THEN  &  GOTO XS,ZS: PRINT "Y"
530  IF A = 15 THEN  &  GOTO XS,ZS: PRINT "Z"
540  GOTO 330
1900  REM
                    ***** ROTATION SUBROUTINE *****


2000 XO = X - XV: REM            TRANSLATION OF COORDINATES TO MOVE
OBSERVER TO ORIGIN OF COORDINATES
2010 YO = Y - YV
2020 ZO = Z - ZV
2025  REM         ***** APPLY ROTATION MATRIX ****
2030 X3 = L1 * XO + L2 * YO + L3 * ZO
2040 Y3 = L4 * XO + L5 * YO + L6 * ZO
2050 Z3 = L7 * XO + L8 * YO + L9 * ZO
2060  REM     ***** PROJECT ROTATED OBJECT ON SCREEN *****
2070 XS = 280 + 40 * D * Y3 / ( - X3)
2080 ZS = 96 - 20 * D * Z3 / ( - X3)
2090  RETURN
3000  REM
                  ***** DATA FOR SCREEN DISPLAY *****

3010  DATA    1,3,3,0,-3,3,0
3020  DATA    2,3,2,0,-3,2,0
3030  DATA    3,3,1,0,-3,1,0
3040  DATA    4,4,0,0,-4,0,0
3050  DATA    5,3,-1,0,-3,-1,0
3060  DATA    6,3,-2,0,-3,-2,0
3070  DATA    7,3,-3,0,-3,-3,0
3080  DATA    8,3,3,0,3,-3,0
3090  DATA    9,2,3,0,2,-3,0
3100  DATA    10,1,3,0,1,-3,0
3110  DATA    11,0,4,0,0,-4,0
3120  DATA    12,-1,3,0,-1,-3,0
```

```
3130    DATA    13,-2,3,0,-2,-3,0
3140    DATA    14,-3,3,0,-3,-3,0
3150    DATA    15,0,0,4,0,0,0
3160    DATA    16,0,0,0,0,0,0
5000    END
```

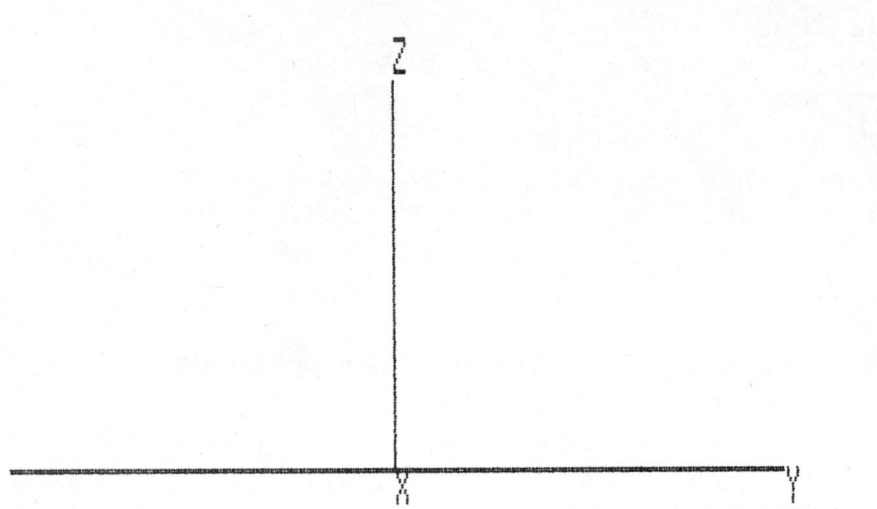

Figure 2.5 In this figure, all of the values of the Eulerian angles are set to zero. The figure is viewed as if the observer is situated on the x-axis at a distance of 20 units from the screen.

Figures plotted with respect to the coordinate system defined in Program 2.4 can be viewed in different orientations by specifying the desired values of the Eulerian angles. Program 2.5 plots the following set of parametric equations and projects the desired view of the figure onto the computer screen.

$$x = 3\sin(3\varphi)\cos(\varphi) \qquad (2.5)$$

$$y = 3\sin(3\varphi)\sin(\varphi) \qquad (2.6)$$

$$z = 3\cos^2(3\varphi) \qquad (2.7)$$

Equation 2.5 and Eq. 2.6 are the parametric form of the equations for the "three leaved rose" in the x-y plane (See Eq. 2.1). Equation 2.7 moves the "leaves" of the figure out of the X-Y plane. The resulting figure is a perspective drawing of a "three-dimensional three leaved rose" superimposed on the coordinate grid defined in the preceding program.

Program 2.5 creates the figure by first forming the coordinate grid defined in the data statements. As before, the rotation subroutine beginning at line number 2000 calculates the location of the vertices

2.3 ROTATION OF COORDINATES IN THREE DIMENSIONS

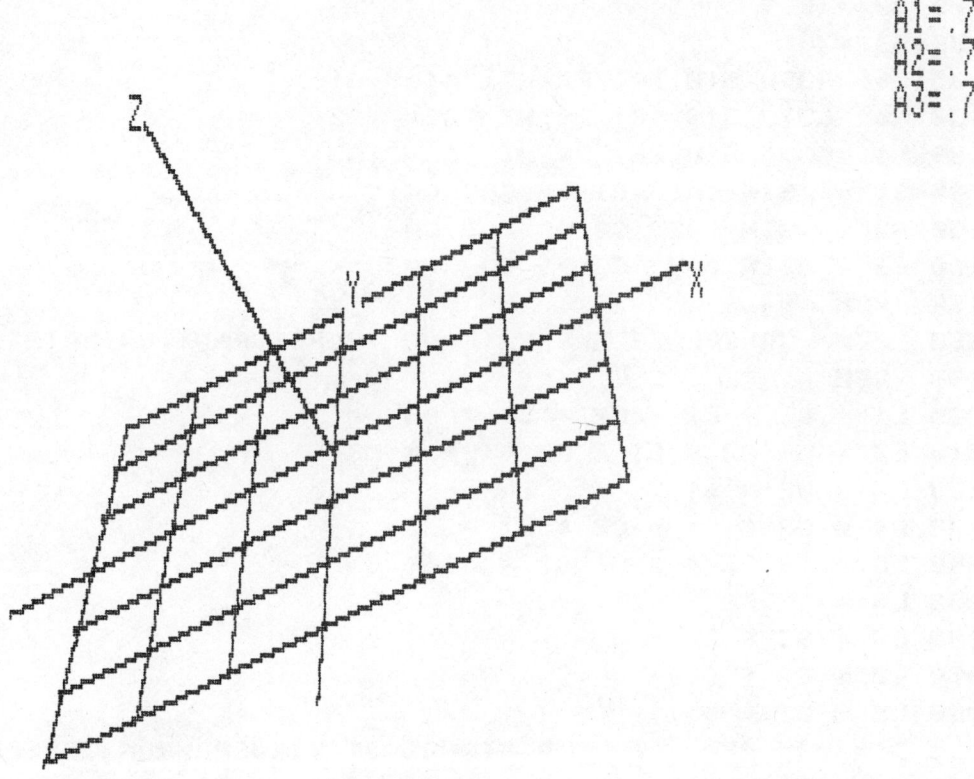

Figure 2.6 To generate this view of the figure, the system is rotated so that the value of each Eulerian angle is 0.7 radians.

of the coordinate grid in the rotated coordinate system and projects the locations onto the two-dimensional screen. Using Eq. 2.5, Eq. 2.6, and Eq. 2.7, values of the coordinates of the figure (x, y, z) are then determined in line numbers 1020-1040. The rotation subroutine is again used to project the points onto the correct screen positions (XS, YS). The points are thus displayed with the same orientation as the coordinate grid formed earlier.

```
7   REM     PROGRAM (2.5)
10  REM   THIS PROGRAM USES EULER'S ANGLES FOR ROTATION
80  REM
                ***** SET UP GRAPHICS CHARACTERISTICS *****


90  & HGR2 : & HCOLOR= 15: & B COLOR= 0: &  PRINT : & MODE(1)
95  REM
                     ***** SET UP SCREEN DISPLAY *****

100 REM       ANGLE A1 IS PHI:REM ANGLE A2 IS THETA:REM ANGLE A3 IS
PSI:REM ALL ANGLES ARE EXPRESSED IN RADIANS
```

```
110 A1 = 0
120 A2 = 0
130 A3 = 0
135  &  GOTO 510,10: PRINT "A1=";A1
137  &  GOTO 510,18: PRINT "A2=";A2
139  &  GOTO 510,26: PRINT "A3=";A3
140 S1 =  SIN (A1):C1 =  COS (A1)
150 S2 =  SIN (A2):C2 =  COS (A2)
160 S3 =  SIN (A3):C3 =  COS (A3)
170  REM                                                    L1  L2  L3
180  REM  CALCULATE ELEMENTS OF TRANSFORMATION MATRIX  L4  L5  L6
190  REM                                                    L7  L8  L9
200 L1 = C3 * C1 - C2 * S1 * S3
210 L2 =  - S3 * C1 - C2 * S1 * C3
220 L3 = S2 * S1
230 L4 = C3 * S1 + C2 * C1 * S3
240 L5 =  - S3 * S1 + C2 * C1 * C3
250 L6 =  - S2 * C1
260 L7 = S3 * S2
270 L8 = C3 * S2
280 L9 = C2
290 D = 20: REM          DISTANCE OF VIEWER FROM SCREEN (CM)
300 XV = D * L1
310 YV = D * L2
320 ZV = D * L3
330  READ A,X1,Y1,Z1,X,Y,Z
340  IF A = 16 GOTO 1000
350  GOSUB 2000
490 SX = XS:SZ = ZS:X = X1:Y = Y1:Z = Z1
495  GOSUB 2000
500  &  HPLOT XS,ZS TO SX,SZ
510  IF A = 4 THEN  &  GOTO XS,ZS: PRINT "X"
520  IF A = 11 THEN  &  GOTO XS,ZS: PRINT "Y"
530  IF A = 15 THEN  &  GOTO XS,ZS: PRINT "Z"
540  GOTO 330
1000  REM
                ***** CALCULATE VALUES AND PLOT FUNCTION *****
1010  FOR PH = 0 TO 3.14 STEP .01
1020 X = 3 *  COS (3 * PH) *  COS (PH)
1030 Y = 3 *  COS (3 * PH) *  SIN (PH)
1040 Z = 3 * ( COS (3 * PH) ^ 2)
1050  GOSUB 2000
1060  &  HPLOT XS,ZS
1070  NEXT PH
```

2.3 ROTATION OF COORDINATES IN THREE DIMENSIONS

```
1080    GOTO 5000
1900    REM             ***** ROTATION SUBROUTINE *****


2000    X0 = X - XV: REM        TRANSLATION OF COORDINATES TO MOVE
OBSERVER TO ORIGIN OF COORDINATES
2010    Y0 = Y - YV
2020    Z0 = Z - ZV
2025    REM             ***** APPLY ROTATION MATRIX ****
2030    X3 = L1 * X0 + L2 * Y0 + L3 * Z0
2040    Y3 = L4 * X0 + L5 * Y0 + L6 * Z0
2050    Z3 = L7 * X0 + L8 * Y0 + L9 * Z0
2060    REM      ***** PROJECT ROTATED OBJECT ON SCREEN *****
2070    XS = 280 + 40 * D * Y3 / ( - X3)
2080    ZS = 96 - 20 * D * Z3 / ( - X3)
2090    RETURN
3000    REM             ***** DATA FOR SCREEN DISPLAY *****

3010    DATA    1,3,3,0,-3,3,0
3020    DATA    2,3,2,0,-3,2,0
3030    DATA    3,3,1,0,-3,1,0
3040    DATA    4,4,0,0,-4,0,0
3050    DATA    5,3,-1,0,-3,-1,0
3060    DATA    6,3,-2,0,-3,-2,0
3070    DATA    7,3,-3,0,-3,-3,0
3080    DATA    8,3,3,0,3,-3,0
3090    DATA    9,2,3,0,2,-3,0
3100    DATA    10,1,3,0,1,-3,0
3110    DATA    11,0,4,0,0,-4,0
3120    DATA    12,-1,3,0,-1,-3,0
3130    DATA    13,-2,3,0,-2,-3,0
3140    DATA    14,-3,3,0,-3,-3,0
3150    DATA    15,0,0,4,0,0,0
3160    DATA    16,0,0,0,0,0,0
5000    END
```

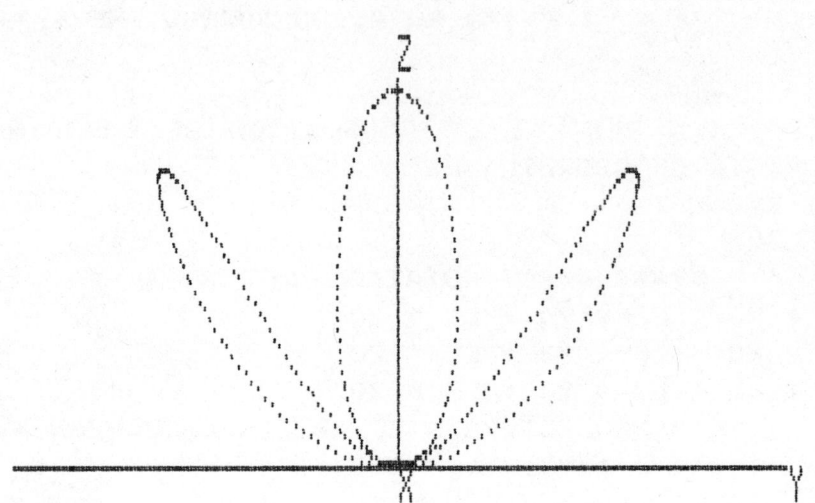

Figure 2.7 In this figure, the value of each Eulerian angle is zero. The figure is viewed as if the observer is situated on the x-axis at a distance of 20 units from the screen.

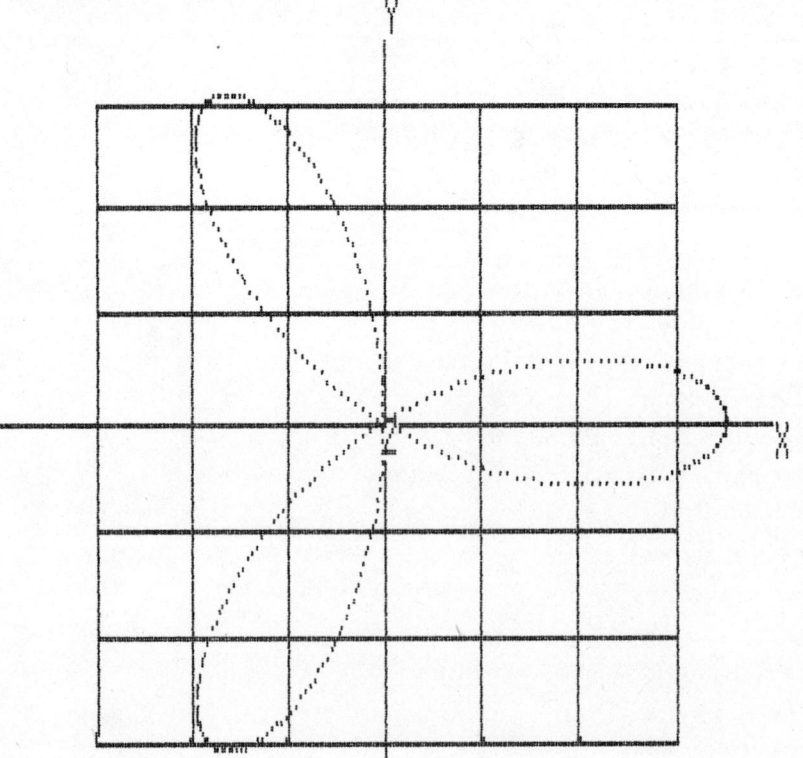

Figure 2.8 The figure is rotated so that the observer is situated on the z-axis. The values of the Eulerian angles are; φ=1.57 radians, θ=1.57 radians, ψ=0 radians.

2.3 ROTATION OF COORDINATES IN THREE DIMENSIONS

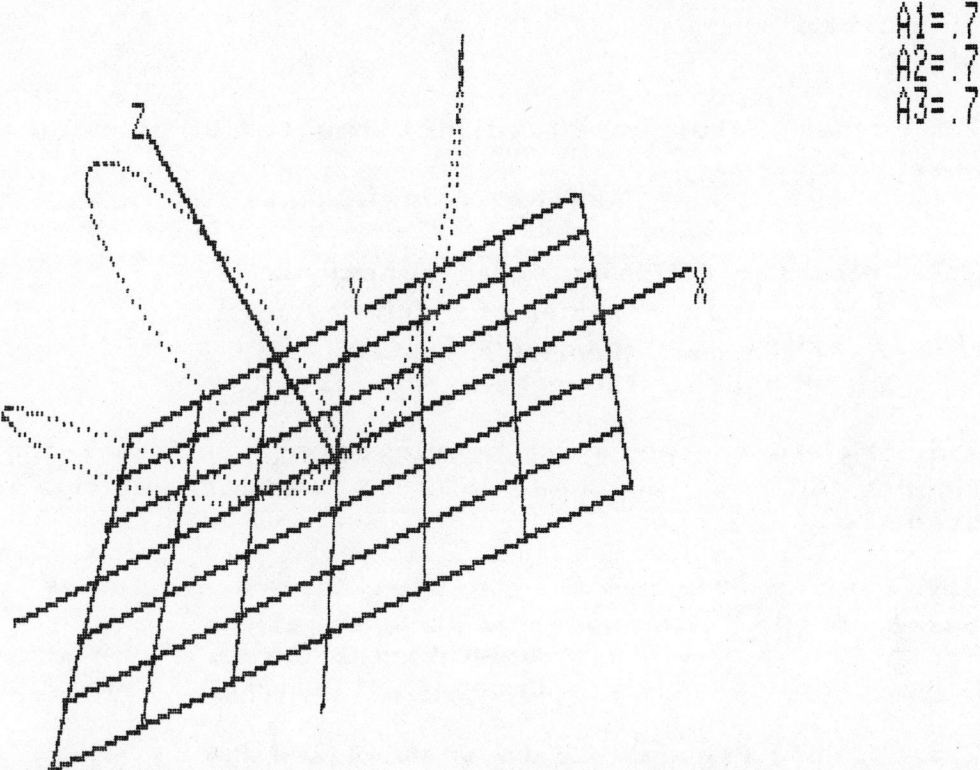

Figure 2.9 To generate this view of the figure, the system is rotated so that the value of each Eulerian angle is 0.7 radians.

The techniques described in this chapter can be used to produce perspective drawings of motion in three dimensions using numerical solutions of the equations of motion to generate a series of points representing the positions of moving objects. These points are then projected onto the computer screen to provide any specified view of the motion. The physicist can thus use the computer screen as a "three-dimensional blackboard" to illustrate the motion of complex systems.

Exercises

2.1. Modify Program 2.1 to plot the family of polar functions (called rose functions):
$$r = r_0 \sin(n\theta) ; \quad n=1,2,3,...$$

2.2. Plot the following polar functions.
  a. $r = a\theta$; (spiral of Archimedes)
  b. $r^2 = b\theta$; (Fermat's spiral)
  c. $r = a/\theta$; (hyperbolic spiral)

2.3. Modify the programs developed in exercises 1 and 2 to rotate the figures through an angle of -1.0 radians relative to the computer screen.

2.4. Alter Program 2.3 to plot Lissajous curves (Bowditch curves) based on the following parametric equations.
$$x = A\cos(nt) + B\sin(nt); \quad n=1,2,3,...$$
$$y = C\cos(mt) + D\sin(mt); \quad m=1,2,3,...$$

2.5. Modify Program 2.3 to draw Figure 2.4.

2.6. Find the values of the Eulerian angles (φ, θ, and ψ) which will cause the x axis of Figure 2.5 to be directed into the screen. (Note: the y axis will be directed toward the left.)

2.7. Find the value of the Eulerian angles which will cause the x-y plane to be oriented so that the viewer is below the plane (on the -z axis).

2.8. Change the viewing distance (D) in Program 2.4 for a few values between 10 cm and 100 cm. Describe the effect of this change on the figure when the Eulerian angles each has a value of 0.7 radians as in Figure 2.6.

2.9. Using a set of parametric equations analogous to those of problem 4, modify Program 2.5 to draw "three-dimensional Lissajous curves."

Problems

2.1. In terms of the usual designation of the x, y, and z axes, the Eulerian angles (φ, θ, ψ) can be described as an initial rotation about the z axis (through angle φ), followed by a rotation about the (rotated) x axis (through angle θ), followed by a final rotation about the rotated z axis (through angle ψ). (See Marion and Thornton, Section 10.7.)

Sets of angles other than the Eulerian angles can be used to describe rotations in three dimensions. A set of angles which is often used in flight simulators consists of a rotation about the z axis followed by a rotation about the (rotated) y axis followed by a final rotation about the (rotated) x axis. These rotation angles are called heading (h), pitch (p), and bank (b) respectively.

a. Find the transformation matrix for rotations described in terms of this set of angles.

b. Write a program similar to Program 2.5 using this set of angles to describe the rotations of the coordinate system. (Hint: See Pickholtz, *Byte*, November, 1982.)

CHAPTER 3

Euler's Method and Associated Numerical Techniques

3.1 Introduction

Many specialized analytical methods have been developed to provide solutions to certain classes of differential equations. Each analytical method works well for a specific class of equations but a new method must be mastered whenever another type of equation is encountered. (See Marion and Thornton, Appendix C.) Euler's method is a universal numerical technique which can be used to generate a numerical solution for any ordinary differential equation or set of coupled ordinary differential equations.

For one-dimensional motion, Newton's Second Law can be written as a second order differential equation of the general form:

$$d^2x/dt^2 = F(x,\dot{x},t)/m \qquad (3.1)$$

In this case, the applied force can depend on position (x), velocity ($\dot{x}$), and time (t).

In more complex cases, application of Newton's Second Law results in sets of equations. When applied to motion of a particle in three dimensions, Newton's Second Law results in a set of three second order differential equations. The equations of motion of systems of interacting particles result in sets of differential equations similar in form to those described above. In many cases, these sets of equations contain coupled differential equations which defy solution in closed form in all but the simplest cases.

3.1 INTRODUCTION

In addition to straightforward application of Newton's Second Law, many problems in classical mechanics require the solution of other second-order differential equations such as those describing the trajectory of particles. In two-dimensional polar coordinates, such an equation might have the general form:

$$d^2r/d\theta^2 = f(r,\theta) \qquad (3.2)$$

Numerical solutions can always be generated for the differential equations of classical mechanics when simple analytical techniques fail. Any computer can be used to perform the repetitive calculations required. By using a computer with high-resolution graphics capability, the numerical results of Euler's method may be expressed graphically to provide a conceptual understanding of the solution. Once a general calculating scheme is developed, the equations and their initial conditions can easily be varied. Numerical "what if" experiments can be performed by making changes in the equations and allowing the computer to repeat the calculations under different sets of conditions.

Euler's method is conceptually straightforward when applied to problems in classical mechanics. Other systems used for numerical solutions of differential equations are more difficult conceptually but often produce more accurate results or increase the efficiency of calculation. Because of the conceptual advantages, only Euler's method and two variations, the half-step approximation and the last point approximation, will be described here.

3.2 Euler's Method

As an example of the general application of Euler's method consider Eq. 3.1, the general form of the second-order differential equation resulting from Newton's Second Law.

Applying the definition of a second derivative, Eq. 3.1 becomes:

$$\lim_{dt \to 0} [dx/dt)_{t+dt} - dx/dt)_t]/dt = F(x,\dot{x},t)/m \qquad (3.3)$$

In Eq. 3.2, $dx/dt)_{t+dt}$ represents the velocity evaluated at the time $t+dt$ and $dx/dt)_t$ represents the velocity evaluated at the time t. It is important to note that this definition is exact only in the limit $dt \to 0$. However, the equation will be approximately correct (to first order) for small finite values of dt. The degree of accuracy of the equation can always be increased for finite values of dt by making the value of dt smaller.

For finite values of dt, Eq. 3.3 becomes:

$$dx/dt)_{t+dt} = dx/dt)_t + [F(x,\dot{x},t)/m]dt \qquad (3.4)$$

Using this equation, the velocity at the time $t+dt$ can be determined from knowledge of the velocity at time t.

From an intuitive point of view, Eq. 3.4 is nothing more than the elementary result for motion with a constant acceleration:

$$v = v_0 + at \qquad (3.5)$$

In Eq. 3.4 the time interval t of Eq. 3.5 is considered to be a short time increment dt. The acceleration during the time increment is considered to be equal to the acceleration at the beginning of the increment: $F(x,\dot{x},t)/m)_t$. During this short time increment the acceleration is considered to be (temporarily) constant so that Eq. 3.4 becomes equivalent to Eq. 3.5. The degree to which this assumption is correct largely determines the accuracy of the calculated value of $dx/dt)_{t+dt}$ for finite time increments, dt.

With knowledge of the velocity, the position at the instant t+dt ($x)_{t+dt}$) can be determined in terms of the position at the instant t ($x)_t$). The definition of a derivative can be used to relate velocity and position:

$$\lim_{dt \to 0} (x)_{t+dt} - x)_t)/dt = dx/dt \qquad (3.6)$$

For finite values of dt, Eq. 3.6 becomes:

$$x)_{t+dt} = x)_t + (dx/dt)_t)dt \qquad (3.7)$$

Equation 3.7 is equivalent to the elementary result: $x=x_0+vt$. The time t is replaced with the time increment dt of Eq. 3.7. The velocity (v) during the time interval is equal to the velocity at the beginning of the time increment $dx/dt)_t$. Although Eq. 3.7 is exactly correct only for motion with a constant velocity, the position of the particle can be determined to any finite degree of accuracy if the time increment dt is made small enough.

If the initial value of velocity and position are known, the value of the acceleration ($F(x,\dot{x},t)/m$) can be calculated at t=0. The velocity at time t=dt can then be calculated from Eq. 3.4.

$$dx/dt)_{t=dt} = dx/dt)_{t=0} + (F(x,\dot{x},t)/m)_{t=0})dt \qquad (3.8)$$

In addition, a new value of the position x can be calculated using Eq. 3.7.

$$x)_{t=dt} = x)_{t=0} + (dx/dt)_{t=0})dt \qquad (3.9)$$

From the velocity and position at time dt the velocity and position can be calculated at the time 2dt.

$$dx/dt)_{t=2dt} = dx/dt)_{t=dt} + (F(x,\dot{x},t)/m)_{t=dt})dt$$
$$x)_{t=2dt} = x)_{t=dt} + (dx/dt)_{t=dt})dt$$

By repeating the calculation for increments of time dt, the velocity and position can be calculated for subsequent times; 3dt, 4dt, ... The number of repetitions of the calculation is determined by the time interval of interest and by the size of the time increment dt.

3.3 Half-Step Approximation

The technique described above (Euler's method) is the simplest and most easily applied method for numerical solution of the differential equations of classical mechanics. In some cases, a slight change in the procedure (called the half-step approximation) can greatly improve the accuracy and speed of the calculations with little increase in the complexity of the calculations. If the applied force is independent of velocity ($\dot{x}$), the general scheme of the half-step approximation is the same as that of Euler's method when applied to Eq. 3.1. However, the velocity is first calculated at the time dt/2 instead of at the time dt.

$$dx/dt)_{t=dt/2} = dx/dt)_{t=0} + (F(x,t)/m)_{t=0} dt/2 \qquad (3.10)$$

The velocity at t=dt/2 found using Eq. 3.10 is considered to be constant during the first time increment from t=0 to t=dt. The position at time dt is then:

$$x)_{t=dt} = x)_{t=0} + (dx/dt)_{t=dt/2} dt \qquad (3.11)$$

From the value of the velocity at the time dt/2 ($dx/dt)_{t=dt/2}$) found using Eq. 3.10, the velocity is then calculated at the time 3dt/2. This velocity can then be used to determine the position at the time 2dt ($x)_{t=2dt}$) using Eq. 3.11.

$$dx/dt)_{t=3dt/2} = dx/dt)_{t=dt/2} + (F(x,t)/m)_{t=dt} dt \qquad (3.12)$$

$$x)_{t=2dt} = x)_{t=dt} + (dx/dt)_{t=3dt/2} dt \qquad (3.13)$$

By repeating the calculation for each time increment dt, the velocity is determined successively at the times; 5dt/2, 7dt/2, ... The position and force are determined for the times; 3dt, 4dt, ... The applied force must be independent of velocity [F=F(x,t)] in order to perform this calculation because the velocity is not calculated at the instants at which the force must be evaluated. The calculation is repeated for each increment of the time interval of interest. Equation 3.14 and Eq. 3.15 are thus the general equations describing the half-step approximation.

$$dx/dt)_{t+dt/2} = dx/dt)_{t-dt/2} + (F(x,t)/m)_{t} dt \qquad (3.14)$$

$$x)_{t+dt} = x)_{t} + (dx/dt)_{t+dt/2} dt \qquad (3.15)$$

The greater accuracy of the half-step approximation (Eq. 3.14 and Eq. 3.15) results in part from considering the average speed for a time increment to occur at the center of the time increment. For uniformly accelerated motion, the average velocity during a time increment is exactly equal to the velocity at the center of the time increment. Because the time increments used in the calculations are small, the acceleration is nearly uniform during the time increment considered. Using the value of the velocity at the center of the time increment (half-step approximation) rather than the value of the velocity at the beginning of the time increment (Euler's method) results in a more accurate calculation due to the slightly more realistic description of accelerated motion.

3.4 Last Point Approximation

In addition to the techniques described above, a simple variation of Euler's method called the last point approximation is very useful in practice. Equation 3.16 and Eq. 3.17 are used to perform the successive calculations of velocity and position. Note that the velocity found for the end of the time increment ($dx/dt)_{t+dt}$) is used to calculate the new position at the end of the increment ($x)_{t+dt}$). Using this variation, the accumulated errors are bounded for oscillatory systems and the resulting solutions are stable without the continuous accumulation of error characterizing the conventional Euler's method. Although this technique does not follow directly from the definition of the derivative or from physical intuition, in practice the results are often better than either of the techniques described earlier. (See Cromer, *American Journal of Physics*, May, 1981.)

$$dx/dt)_{t+dt} = dx/dt)_t + (F(x,\dot{x},t)/m)_t)dt \qquad (3.16)$$

$$x)_{t+dt} = x)_t + (dx/dt)_{t+dt})dt \qquad (3.17)$$

For second-order differential equations involving variables other than position and time, Euler's method and its variations can be applied in a manner similar to that described above. The dependent variable in the differential equation is substituted for position (x) and the independent variable replaces time (t). A numerical solution can then be generated by repetitive evaluation of equations analogous to Eq. 3.4 and Eq. 3.5 (Euler's method); Eq. 3.14 and Eq. 3.15 (half-step approximation); or Eq. 3.16 and Eq. 3.17 (last point approximation).

The choice of method depends on the problem being solved. Euler's method is particularly simple both conceptually and in terms of development of computer programs. However, (especially in the case of bound systems) the last point approximation will often greatly improve the accuracy of the solution without slowing the operation of the program but with some loss of intuitive understanding. Students may be able to develop their own variations combining these three methods to provide a rapid solution to a particular problem. In practice, it is useful to compare the methods by predicting some dynamic variable of the system (such as energy) and testing the methods to determine which method produces the smallest deviations from the ideal value of the variable. If a highly accuracte solution is

3.3 LAST POINT APPROXIMATION

required, the fourth order Runge-Kutta method described in Appendix C can be used.

The following programs make use of the three methods presented above. Clarity, accuracy, and speed of operation were all considered in developing the programs and choosing the preferred method of calculation. The programs are designed to rapidly produce results with pedagogical value. If extremely accurate results are required, the programs will require modification in most cases.

CHAPTER 4

Motion In One Dimension

4.1 Introduction

When beginning to develop computer programs which apply Euler's method to problems in classical mechanics, it is helpful to first compare the results of analytical and numerical methods for solving differential equations. Equation 4.1, the equation of motion describing horizontal motion of a particle in a resisting medium, is used here to provide a comparison of these methods.

$$mdv/dt = -kmv \qquad (4.1)$$

This differential equation can be solved analytically by integrating to find the velocity v of the moving object as a function of time. Integration of the resulting equation is then performed to obtain the position x as a function of time. (See Marion and Thornton, Section 2.4, Example 2.4.) If the initial conditions of the motion at $t=0$ are $v=v_0$ and $x=0$, this technique yields the following equations for velocity and position as a function of time.

$$v = v_0 \exp(-kt) \qquad (4.2)$$

$$x = v_0[1 - \exp(-kt)]/k \qquad (4.3)$$

4.2 Euler's Method

Equation 4.1 can also be solved numerically using Euler's method. Values of velocity and position found numerically can then be compared with the predictions of Eq. 4.2 and Eq. 4.3. In order to apply Euler's method, both sides of Eq. 4.1 are first divided by m followed by application of Eq. 3.8 and Eq. 3.9. This yields:

$$dx/dt)_{t+dt} = dx/dt)_t + (-kv)_t)dt \qquad (4.4)$$

$$x)_{t+dt} = x)_t + (dx/dt)_t)dt \qquad (4.5)$$

4.3 Computer Program

The BASIC language computer program below uses Eq. 4.4 and Eq. 4.5 to calculate successive values of velocity and position as a function of time for the object whose motion is described by Eq. 4.1. The resulting values of velocity and position after an elapsed time of 10 seconds are printed on the screen. The program is later altered to compare the results of the numerical calculation with the values of velocity and position found using Eq. 4.2 and Eq. 4.3 which were derived using analytical methods.

```
100 V=10   :REM INITIAL VELOCITY (METERS/SECOND)
110 X=0    :REM INITIAL POSITION (METERS)
120 K=2    :REM FLUID FRICTION COEFFICIENT (SEE EQ. 4.3)
130 DT=.1
140 FOR T=0 TO 10 STEP DT
150 V1=V-K*V*DT  :REM EQ. 4.4
160 X1=X+V*DT    :REM EQ. 4.5
170 V=V1
180 X=X1
190 NEXT T
200 REM PRINT VALUES OF VELOCITY AND POSITION WHEN T=10 SECONDS
210 PRINT "T=";T;" SECONDS"
220 PRINT "X=";X;" METERS"
230 PRINT "V=";V;" METERS/SECOND"
```

In the program above, line number 130 specifies the time increment dt between successive evaluations of position and velocity. Line numbers 170-180 redefine the variables V and X ($dx/dt)_t$ and $x)_t$) in order to use the newly determined values V1 and X1 ($dx/dt)_{t+dt}$ and $x)_{t+dt}$) for the values of velocity and position at the beginning of the time increment in the next repetition of the calculation. The values of velocity and position are printed after the time of 10 seconds specified in line number 140 has elapsed.

The following program lines can be added to the program above in order to compare the results of numerical calculations using Euler's method with the analytical results of Eq. 4.2 and Eq. 4.3.

```
240 REM PRINT VALUES OF V AND T AFTER 10 SECONDS CALCULATED FROM
EQ. 4.2 AND EQ. 4.3
250 V=10*EXP(-K*T)          :REM EQ. 4.2
260 X=10*(1-EXP(-K*T))/K    :REM EQ. 4.3
270 PRINT "X=";X;" METERS"
280 PRINT "V=";V;" METERS/SECOND"
```

The values of the variables X and V printed after execution of line numbers 240-280 are based on exact analytical results instead of approximate numerical calculations. Errors caused by the finite size of the time increment dt and by the limited accuracy of the approximations used in the numerical calculations cause most of the difference in the results. A difference of this type is called truncation error.

The relative agreement of the final values of position and velocity depends on the size of the time increment dt between successive calculations in the first part of the program. The program can be modified to use both smaller and larger values of dt in order to study the effect of the size of the time increment on the results of the calculations.

In principle, smaller values of dt yield more accurate final values of position and velocity. In practice, however, the use of extremely small values of dt may decrease accuracy because of the increased number of successive calculations required to produce the results. Error of this type is due to the limited number of significant figures used in computer calculations and is called roundoff error. The use of double precision calculations reduces roundoff errors but is not necessary for accurate results with the programs of this text.

The values found using the program developed above also depend on the type of computer used to perform the calculations. The difference in results found using various computers is due to the different numbers of significant figures which the machines use to perform calculations. Table 4.1 illustates the results found using several different computers to perform the calculations using the program above. The lines for which dt is not specified refer to the results calculated using Eq. 4.2 and Eq. 4.3.

In practice, truncation error is the largest source of error in calculations which apply the methods described in this text. The magnitude of truncation error is often limited when using the half-step approximation or the last point approximation. However, the accumulated error in the calculated energy of a particle is proportional to the magnitude of the forces encountered and to the size of the time increment between calculations. The nature of the motion of the object may appear changed if large forces are encountered. For example, in the case of orbital motion, a near miss of the orbiting particles can cause the calculated energy to change sign from negative (bound particle) to positive (unbound particle). More elaborate, higher order approximations such as the fourth order

Runge-Kutta method described in Appendix C should be applied when it is necessary to deal with problems of this type.

Table (4.1)

APPLE IIe (Applesoft BASIC)

dt(s)	x(m)	v(m/s)
.1	5.00000001	2.03703598E-09
.01	4.99999995	1.68296735E-08
.001	4.99999997	2.01624559E-08
	4.99999999	2.06115363E-08

IBM-pc (GW BASIC, single precision)

dt(s)	x(m)	v(m/s)
.1	5.000000	2.037037E-09
.01	4.999990	1.682951E-08
.001	4.999887	2.020101E-08
	5.000000	2.061145E-08

IBM-pc (GW BASIC, double precision)

dt(s)	x(m)	v(m/s)
.1	4.999999998981482	2.037035976334486D-09
.01	4.999999991585163	1.682967357215956D-08
.001	4.999999989898563	2.0202860902385D-08
	4.999999989694233	2.061153470123145D-08

DEC PDP 11-70 (DEC BASIC)

dt(s)	x(m)	v(m/s)
.1	4.99999999898148	2.03703597633449E-09
.01	4.99999999158516	1.68296735721596E-08
	4.99999998969423	2.06115362243856E-08

The program developed above calculates and prints the values of position and velocity after the time specified in the for-next loop has elapsed. This program is the basis for the development of programs which graphically display the successive values of position and velocity.

Line numbers 170, 180, and 190 are first renumbered to move the lines to the end of the program. (Line numbers 290, 300, and 310 are used in the example below.) When this program is executed, the calculated values of V and X are printed for each time increment dt. As the values are printed, the screen is scrolled with the printing of each new line. As a first step toward a graphical display of these calculations insert a new program line: 205 HOME. For the Apple IIe, execution of this statement causes the screen to be cleared each time a new set of values is printed and establishes a constantly changing display of the values of velocity and position. These printed values can now be scaled and plotted graphically on the computer screen.

4.4 Graphics Program

In order to convert this program to a graphics program, program line numbers 200 through 280 are deleted. (If using Applesoft BASIC, enter and execute the command line DEL 200,280.) This newly created gap in the line numbers will be the location of program lines used to produce a graphic display of the calculated values of position and velocity plotted as a function of time.

Insertion of the program lines below into the program alters the program to generate a graph of the velocity as a function of time for an object moving in accordance with Eq. 4.1. Velocity is plotted on the vertical axis and elapsed time is plotted on the horizontal axis of the graph.

```
100 &HGR2:&HCOLOR=15:&MODE(1):&PRINT

200 TS= 50 + 50*T    :REM TS IS HORIZONTAL SCREEN POSITION
210 VS= 90 - 9*V     :REM VS IS VERTICAL SCREEN POSITION
220 &HPLOT TS,VS
```

This program produces an unlabelled curve representing the variation of velocity (plotted vertically) as a function of time (plotted horizontally). Inserting the following program lines adds a graphical display of the position (plotted vertically) as a function of time. The graphs are produced simultaneously as the calculations proceed.

```
230 XS= 180 - 10*X   :REM XS IS VERTICAL SCREEN POSITION
240 &HPLOT TS,XS
```

4.5 Scaling the Figure

The double-resoloution mode of the Apple IIe has a screen width of 560 pixels. The statement in line number 200 positions the horizontal axis such that the time t=0 corresponds to a position 50 pixels from the left edge of the screen. In addition, the graph is scaled so that an elapsed time of one second corresponds to an additional horizontal displacement of 50 pixels.

The screen height for this graphics system is 192 pixels. Program line number 210 positions the vertical axis such that v=0 corresponds to a position 90 pixels from the top of the screen. The scale of the graph is defined so that a velocity increase of 1 meter/second produces an upward displacement of 9 pixels on the computer screen.

The statement in line number 230 then positions a lower graph of position (x) versus time (t) such that x=0 correspond to a screen location 180 pixels from the top of the screen. The screen is scaled so that a forward movement of one meter produces an upward displacemant of 9 pixels on the computer screen.

Students should experiment with line numbers 200-230 by changing the factors which control the position and size of the graphs. In addition, the time interval over which the motion is evaluated and the time increment (dt) between calculations should also be changed (line

4.5 SCALING THE FIGURE

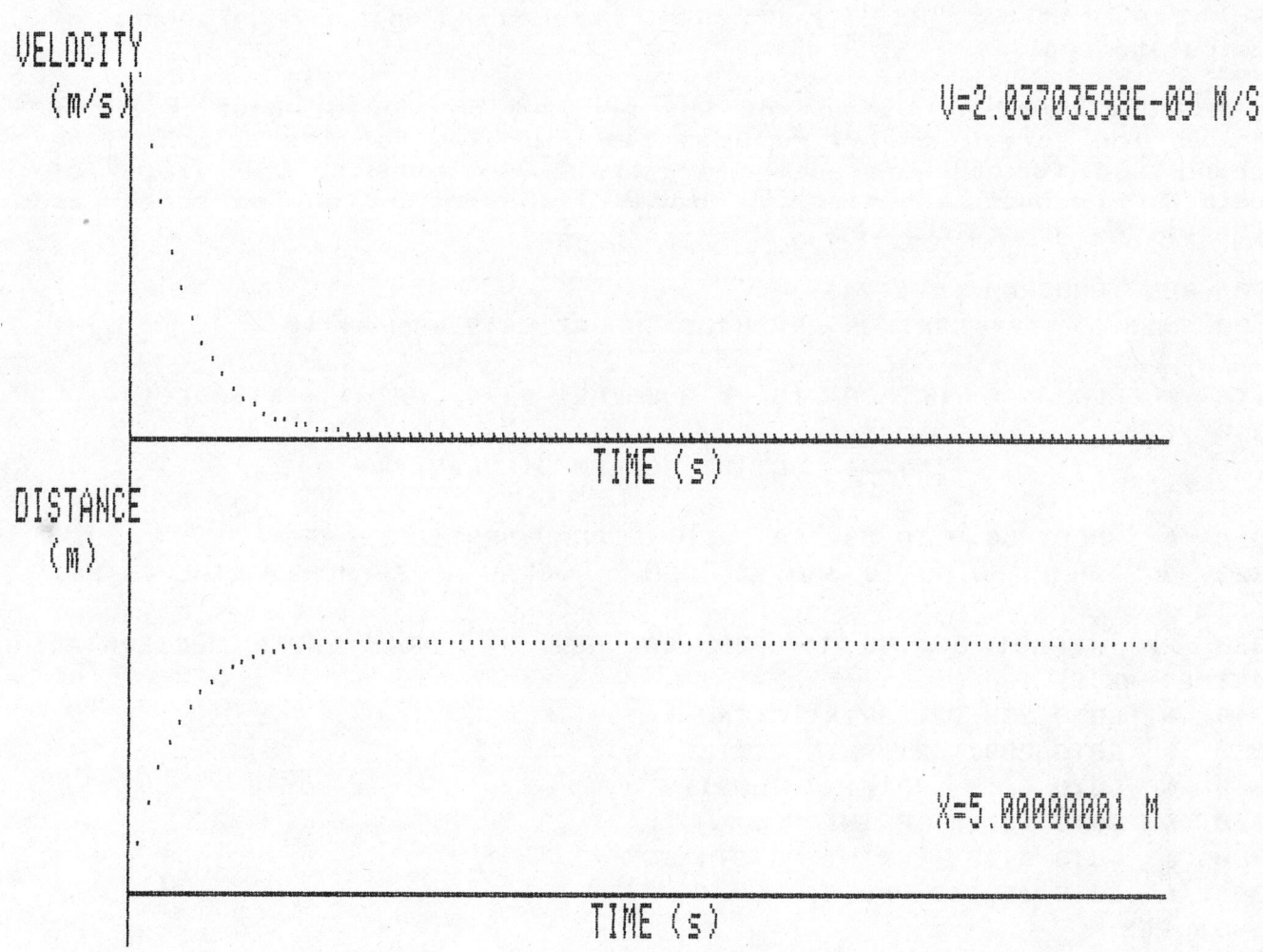

Figure 4.1. This figure displays the velocity-time curve and the position-time curve plotted using the initial conditions indicated in Program 4.1.

number 140). The initial conditions and the mechanical description of the force can also be altered in order to study how these variables affect the motion of the object.

The program which was developed in the preceding sections illustrates the general outline of computer programs which graphically display the results of solutions to problems in classical mechanics using Euler's method. The general pattern of these programs is outlined below.

1. Set-up for graphics characteristics.
2. Establish screen display.
3. Specify initial conditions, time increment, and other variables related to the mechanical description of the system.

4. Repetitive calculations based on either Eq. 3.4 and Eq. 3.5 (Euler's method) or Eq. 3.14 and Eq. 3.15 (half-step approximation) or Eq. 3.16 and Eq. 3.17 (last point approximation) and graphing of calculated values.

A more complete version of the program developed above is shown below. The screen display includes labelled axes for the velocity-time graph and for the distance-time graph. In addition, the values of velocity, position, and time are updated and printed on the screen as the values are calculated.

```
90  REM   PROGRAM (4.1)
100 REM     ***** SET UP GRAPHICS CHARACTERISTICS *****

110 &  HGR2 : &  HCOLOR= 15: &  B COLOR= 0: &  PRINT : &  MODE(1)
300 REM
                  *****   SET UP SCREEN DISPLAY *****

310 &  HPLOT 50,0 TO 50,190: REM      DRAW VERTICAL AXES
320 &   HPLOT 50,90 TO 550,90: REM       DRAW UPPER HORIZONTAL (TIME) AXIS
330 &    HPLOT 50,180 TO 550,180: REM       DRAW LOWER HORIZONTAL (TIME) AXIS
340 &  GOTO 250,92: PRINT "TIME (s)"
350 &  GOTO 250,182: PRINT "TIME (s)"
360 &  GOTO 0,10: PRINT "VELOCITY"
370 &  GOTO 15,20: PRINT "(m/s)"
380 &  GOTO 0,100: PRINT "DISTANCE"
390 &  GOTO 15,110: PRINT "(m)"
500 REM
                ***** SPECIFY INITIAL CONDITIONS *****
510 X = 0
520 V = 10
530 K = 2
540 DT = .1
1000 REM
            *****   CALCULATE AND PLOT VELOCITY AND POSITION *****

1010  FOR T = 0 TO 10 STEP DT
1020  V1 = V - K * V * DT: REM     EQ. 4.4
1030  X1 = X + V * DT: REM      EQ. 4.5
1040  TS = 50 + 50 * T: REM      TS IS  HORIZONTAL SCREEN POSITION TO PLOT TIME
1050  VS = 90 - 9 * V: REM         VS IS VERTICAL SCREEN POSITION TO PLOT VELOCITY
1060 &  HPLOT TS,VS
```

```
1070 XS = 180 - 10 * X: REM      XS IS VERTICAL SCREEN POSITION TO
PLOT POSITION
1080  &  HPLOT TS,XS
1110 X = X1:V = V1
1120  &  GOTO 400,160: PRINT "X=";X;" m   "
1130  &  GOTO 400,20: PRINT "V=";V;" m/s   "
1140 NEXT T
2000 END
```

4.6 Rockets

A rocket is an example of a system whose mass depends on time. Equation 4.7 is the equation of motion for a rocket which burns fuel at a rate $\dot{m}$ with an exhaust velocity V in the absence of external forces. (See Marion and Thornton, Section 2.7.)

$$m d^2x/dt^2 = V\dot{m} \tag{4.7}$$

For many rockets, fuel is burned at a constant rate and $\dot{m}$ is thus a constant. For a rocket of this type with an initial mass m_0, and carrying an additional payload m_p, the initial mass of the system is $m_0 + m_p$. As the fuel is exhausted, the mass m being accelerated at any instant is:

$$m = m_p + m_0 - \dot{m}t \tag{4.8}$$

Substituting Eq. 4.8 for the mass m of the rocket, Eq. 4.7 becomes:

$$d^2x/dt^2 = V\dot{m}/(m_p + m_0 - \dot{m}t) \tag{4.9}$$

Applying Eq. 3.8 this becomes:

$$dx/dt\}_{t+dt} = dx/dt\}_t + [V\dot{m}/(m_p + m_0 - \dot{m}t)]dt \tag{4.10}$$

In Program 4.2, Eq. 4.10 is evaluated to plot a velocity-time graph for the motion of a rocket which burns fuel at a constant rate. The time for which the rocket motor burns depends on the rate $\dot{m}$ at which fuel is burned and on the amount of fuel carried. If (as specified in the program below) the rocket is constructed so that the mass of the fuel is 97 percent of the mass of the rocket, the rocket motor will operate for a time, $t = .97 m_0/\dot{m}$.

```
90  REM   PROGRAM (4.2)
100 REM     ***** SET UP GRAPHICS CHARACTERISTICS *****

110  &  HGR2 :  &  HCOLOR= 15:  &  B COLOR= 0:  &  PRINT :  &  MODE(1)
```

```
300  REM
             ***** SET UP SCREEN DISPLAY *****

310  &  HPLOT 50,0 TO 50,160: REM   DRAW VERTICAL (VELOCITY) AXIS
320  &  HPLOT 50,160 TO 550,160: REM   DRAW HORIZONTAL (TIME) AXIS
330  FOR T = 0 TO 10 STEP 2
340  X = 50 + 50 * T
350  &   GOTO X,160: &  HPLOT X,160 - 3 TO X,160: &   GOTO X - 4,160
 + 3:  PRINT T: REM   LABEL HORIZONTAL AXIS
360  NEXT T
370  &  GOTO 290,175: PRINT "TIME (s)"
380  FOR V = 0 TO 1500 STEP 200
390  Y = 160 - .1 * V
400  &   HPLOT 50,Y TO 50 + 6,Y: &   GOTO 50 - 35,Y: PRINT V: REM
LABEL VERTICAL AXIS
410  NEXT V
420  &  GOTO 10,5: PRINT "VELOCITY (m/s)"
500  REM
             ***** SPECIFY INITIAL CONDITIONS *****

510  V = 0: REM    INITIAL VELOCITY
520  VE = 300: REM     EXHAUST VELOCITY
530  DM = 100: REM     RATE AT WHICH FUEL IS BURNED (DM/DT)
540  MP = 10: REM      MASS OF PAYLOAD
550  M0 = 1000: REM     INITIAL MASS OF ROCKET
560  T1 = .98 * M0 / DM: REM    TIME TO BURN ALL FUEL
570  DT = .1: REM    TIME INCREMENT BETWEEN CALCULATIONS
1000  REM
             ***** CALCULATE AND PLOT VELOCITY *****

1010  FOR T = 0 TO T1 STEP DT
1020  V1 = V + (VE * DM / (M0 + MP - DM * T)) * DT: REM    EQ. 4.10
1030  YS = 160 - .1 * V: REM        VERTICAL SCREEN POSITION TO PLOT V
1040  XS = 50 + 50 * T: REM        HORIZONTAL SCREEN POSITION TO PLOT
T
1050  &  HPLOT XS,YS
1060  V = V1:X = X1
1070  NEXT T
2000  END
```

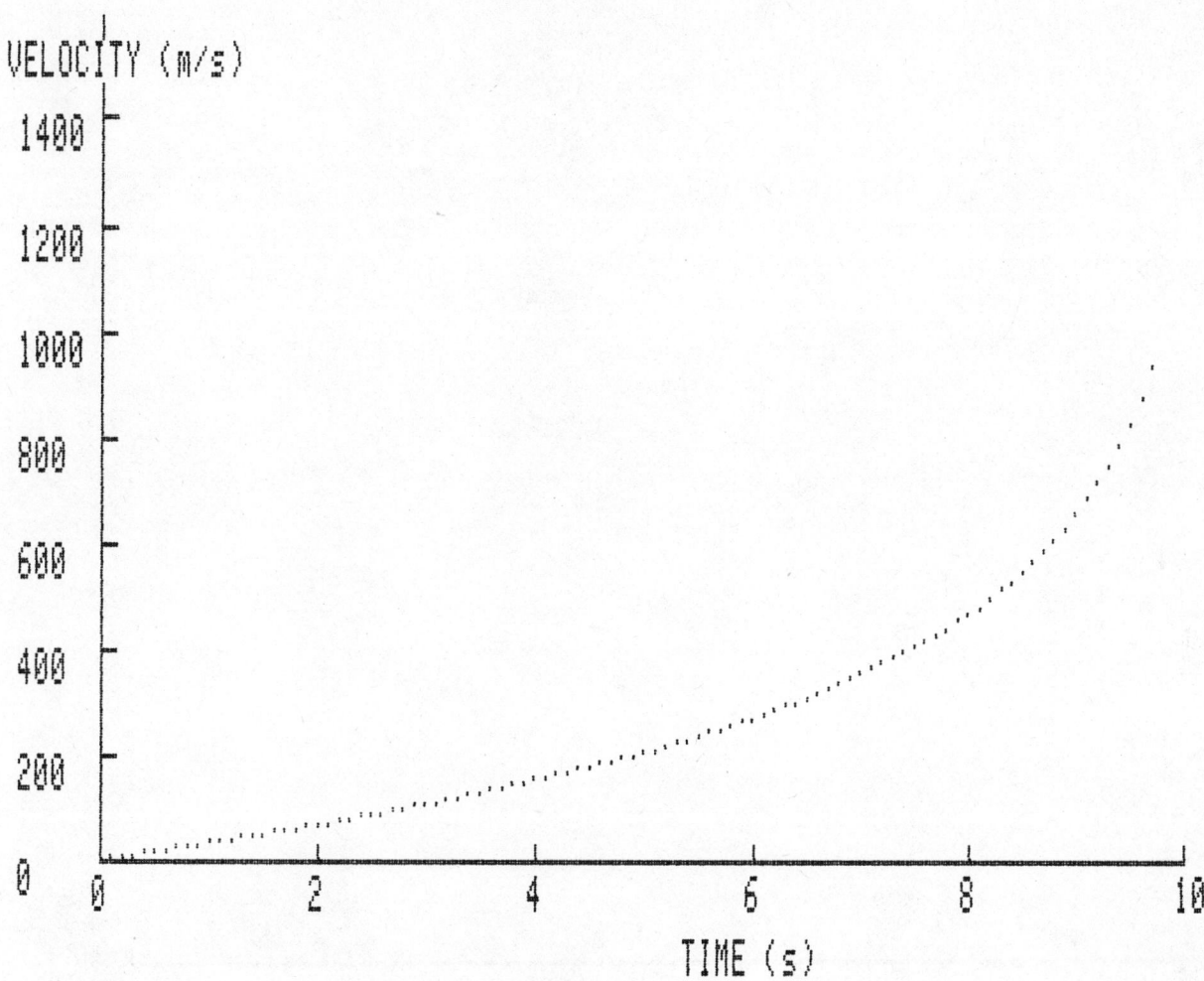

Figure 4.2. This velocity-time graph was drawn using Program 4.2. As the rocket mass decreases due to burning of fuel, the slope of the graph (acceleration) increases.

4.7 Two-Stage Rockets

Program 4.2 can be extended to apply to a two-stage rocket used to accelerate a payload of mass m_p. The initial mass of the first stage is m_{10} and the second stage has an initial mass m_{20}. The rate at which fuel is burned in the rocket motor of the first stage is $\dot{m}_1$. Thus the mass being accelerated at any instant during the operation of the first stage is: $m = m_{10} + m_{20} + m_p - \dot{m}_1 t$.

After the first stage motor has completed its operation, the second stage fires, burning fuel at the rate, $\dot{m}_2$. The mass being accelerated at any instant during the operation of the second stage rocket motor is: $m = m_{20} + m_p - \dot{m}_2 t$.

In Program 4.3 the mass of the fuel is specified to be 95 percent of the initial mass of the first stage. Thus the first stage rocket motor burns for a time, $t = .95 m_{10}/\dot{m}_1$. The mass of the fuel in the

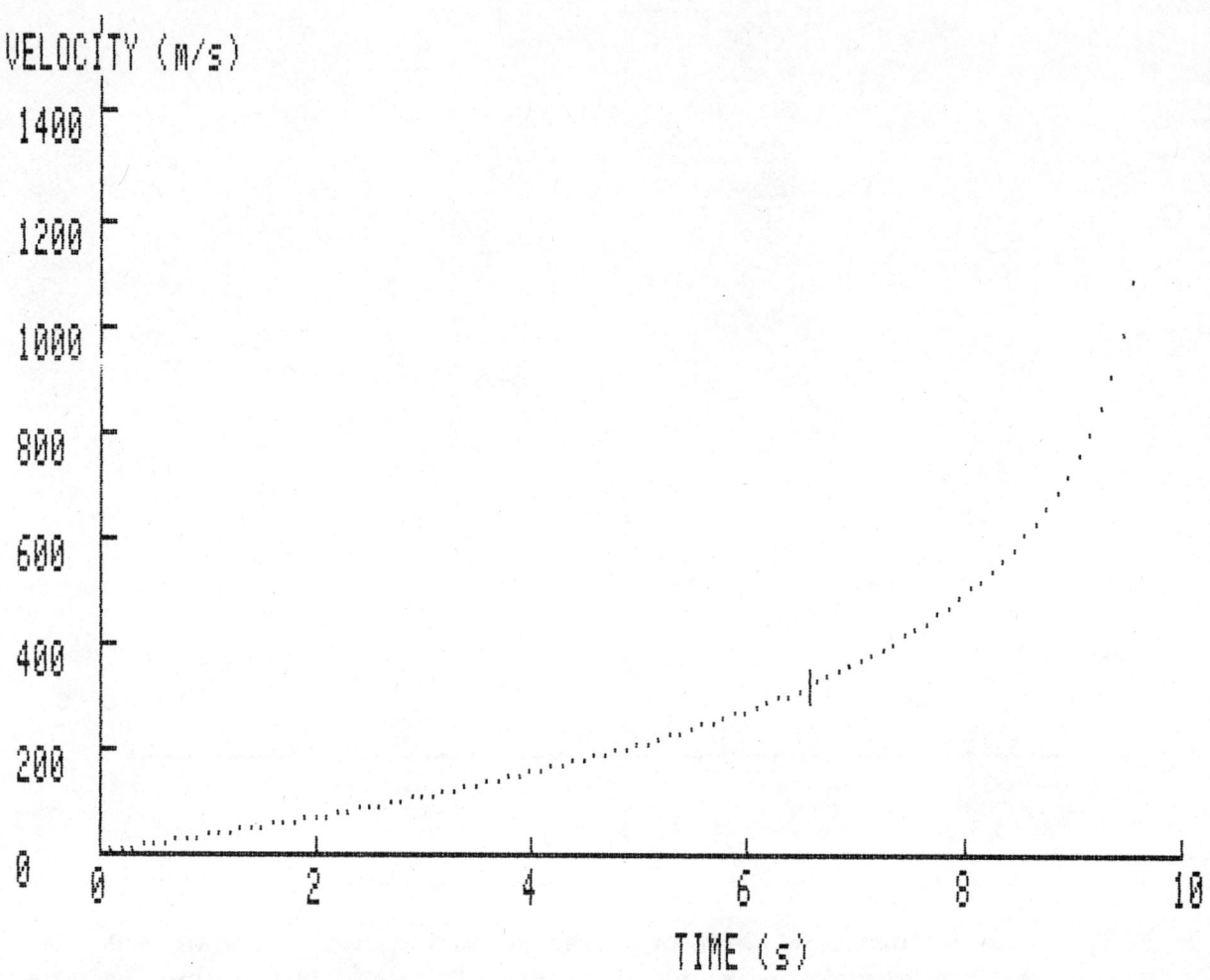

Figure 4.3. The velocity-time graph drawn using Program 4.3 is shown in this figure. Note that the final velocity (of the payload) after all fuel is exhausted is significantly higher for the two-stage rocket than for the single-stage rocket (Figure 4.2) of equal initial mass.

second stage is specified to be 97 percent of the initial mass of the second stage causing the rocket motor of the second stage to operate for a time, $t = .97 m_{20}/\dot{m}_2$.

In order to calculate the velocity of a two-stage rocket, Eq. 4.10 is initially evaluated for the first stage of the rocket until enough time has elapsed to exhaust the fuel in the first stage. After that time, (beginning at line number 1090) the equation of motion for the second stage is evaluated using the final velocity of the first stage as the initial velocity of the second stage. Program 4.3 evaluates Eq. 4.10 under these conditions and graphically displays the results of the calculations. A vertical line is drawn to mark the firing of the second stage (line number 1080).

4.7 TWO-STAGE ROCKETS

```
90   REM   PROGRAM (4.3)
100  REM       ***** SET UP GRAPHICS CHARACTERISTICS *****

110  & HGR2 : & HCOLOR= 15: & B COLOR= 0: & PRINT : & MODE(1)
300  REM
              *****  SET UP SCREEN DISPLAY *****

310  & HPLOT 50,0 TO 50,160: REM  DRAW VERTICAL (VELOCITY) AXIS
320  & HPLOT 50,160 TO 550,160: REM  DRAW HORIZONTAL (TIME) AXIS
330  FOR T = 0 TO 10 STEP 2
340  X = 50 + 50 * T
350  & GOTO X,160: & HPLOT X,160 - 3 TO X,160: & GOTO X - 4,160
     + 3: PRINT T: REM  LABEL HORIZONTAL AXIS
360  NEXT T
370  & GOTO 290,175: PRINT "TIME (s)"
380  FOR V = 0 TO 1500 STEP 200
390  Y = 160 - .1 * V
400  & HPLOT 50,Y TO 50 + 6,Y: & GOTO 50 - 35,Y: PRINT V: REM
     LABEL VERTICAL AXIS
410  NEXT V
420  & GOTO 10,5: PRINT "VELOCITY (m/s)"
500  REM
             ***** SPECIFY INITIAL CONDITIONS *****

510  V = 0: REM     INITIAL VELOCITY
520  VE = 300: REM      EXHAUST VELOCITY (BOTH STAGES)
530  DM = 100: REM       RATE AT WHICH FUEL IS BURNED (DM/DT FOR BOTH
     STAGES)
540  MP = 10: REM       MASS OF PAYLOAD
550  M10 = 700: REM      INITIAL MASS OF FIRST STAGE
560  M20 = 300: REM   INITIAL MASS OF SECOND STAGE
600  T1 = .95 * M10 / DM: REM          TIME TO BURN ALL FUEL IN FIRST
     STAGE
610  T2 = .97 * M20 / DM: REM  TIME TO BURN ALL FUEL IN SECOND STAGE
620  DT = .1: REM     TIME INCREMENT BETWEEN CALCULATIONS
1000 REM
             *****  CALCULATE AND PLOT VELOCITY *****

1010 FOR T = 0 TO T1 STEP DT
1020 V1 = V + (VE * DM / (M10 + M20 + MP - DM * T)) * DT: REM EQ.
     4.10
1030 YS = 160 - .1 * V: REM       VERTICAL SCREEN POSITION TO PLOT V
1040 XS = 50 + 50 * T: REM         HORIZONTAL SCREEN POSITION TO PLOT
     T
1050 & HPLOT XS,YS
```

```
1060  V = V1:X = X1
1070  NEXT T
1080  &  HPLOT XS,YS - 3 TO XS,YS + 3: REM   MARKER FOR SECOND STAGE
IGNITION
1090  FOR T = T1 TO T1 + T2 STEP DT
1100  V1 = V + (VE * DM / (M20 + MP - DM * (T - T1))) * DT
1110  YS = 160 - .1 * V
1120  XS = 50 + 50 * T
1130  &  HPLOT XS,YS
1140  V = V1:X = X1
1150  NEXT T
2000  END
```

4.8 Simple Pendulum

The equation of motion of a simple pendulum is found by considering the the torque acting on the pendulum. The torque acts to return the pendulum toward the equilibrium position and is proportional in magnitude to the sine of the angle of deflection of the pendulum from equilibrium. The resulting equation of motion (Eq. 4.11) is a non-linear second order differential equation which is solved analytically by assuming a solution in the form of an infinite series. (See Marion and Thornton, Section 3.13.)

$$d^2\theta/dt^2 = -g\sin(\theta)/l \qquad (4.11)$$

Equation 4.11 can easily be solved using any of the methods described in Chapter 3. The half-step approximation is used below to solve for the angular position θ of the pendulum as a function of time.

Applying Eq. 3.14 and Eq. 3.15, the equation of motion for the simple pendulum becomes:

$$d\theta/dt\big)_{t+dt/2} = d\theta/dt\big)_{t-dt/2} + (g\sin(\theta)\big)_t/l)dt \qquad (4.12)$$

$$\theta\big)_{t+dt} = \theta\big)_t + (d\theta/dt\big)_{t+dt/2})dt \qquad (4.13)$$

These equations are the basis for Program (4.4) which evaluates the motion of a simple pendulum using the half-step approximation. The angular position θ of the pendulum is plotted as a function of time t for the initial conditions specified in the program. Students should study the motion under different conditions by editing the program to vary the initial values of angular velocity and angular position. In addition, the mechanical description of the pendulum can be modified by changing the length of the pendulum or the assumed value of the acceleration due to gravity.

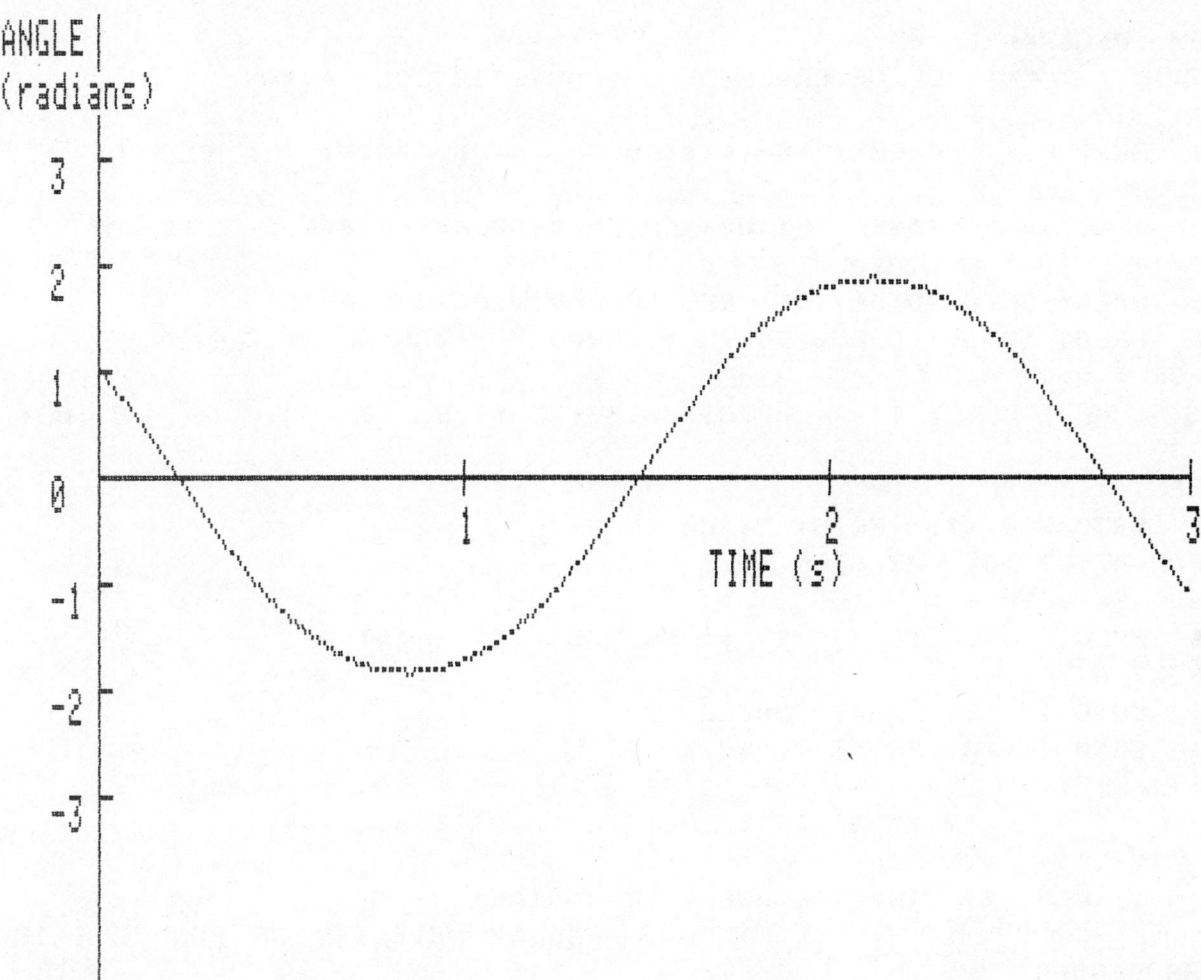

Figure 4.4. This figure shows the angular position as a function of time for a swinging pendulum that was set into motion from an initial angular position of 1 radian and with an initial angular velocity (toward the equilibrium position) of -4 rad/s.

The programs in this chapter can be extended to solve for the motion of many one-dimsional systems. For example, the equation of motion for the damped harmonic oscillator can be solved analytically but a great deal of intuitive understanding can be gained by solving the equation numerically. In addition, complex driving forces and friction forces applied to oscillators can be studied using these techniques. The types of problems subjected to analysis are thus limited only by imagination, not by the limits of analytical mathematics.

4 · MOTION IN ONE DIMENSION

```
90   REM   PROGRAM (4.4)
100  REM     ***** SET UP GRAPHICS CHARACTERISTICS *****

110  & HGR2 : & HCOLOR= 15: & B COLOR= 0: & PRINT : & MODE(1)
300  REM
                 *****   SET UP SCREEN DISPLAY *****

310  & HPLOT 50,0 TO 50,190: REM    DRAW VERTICAL AXIS
320  & HPLOT 50,96 TO 550,96: REM   DRAW HORIZONTAL AXIS
330  FOR T = 1 TO 3
340  XS = 50 + 150 * T: & HPLOT XS,99 TO XS,93: & GOTO XS - 2,102:
PRINT T
350  NEXT T
360  & GOTO 300,110: PRINT "TIME (s)"
370  FOR A = 3 TO  - 3 STEP  - 1
380  YS = 96 - 20 * A
390  & HPLOT 50,YS TO 55,YS: & GOTO 30,YS: PRINT A
400  NEXT A
410  & GOTO 10,10: PRINT "ANGLE"
420  & GOTO 10,20: PRINT "(radians)"
500  REM
                 ***** SPECIFY INITIAL CONDITIONS *****

510  A = 1: REM      INITIAL ANGLE IN RADIANS
520  V =  - 4: REM           INITIAL ANGULAR VELOCITY OF PENDULUM IN
RADIANS/SECOND
530  L = 1: REM   LENGTH OF PENDULUM IN METERS
540  G = 9.8: REM    ACCELERATION DUE TO GRAVITY
550  DT = .01
1000 REM
             ***** CALCULATE AND PLOT ANGULAR  POSITION *****

1010 V = V - (G *  SIN (A) / L) * DT / 2: REM    VELOCITY AT DT/2
1020  FOR T = 0 TO 3 STEP DT
1030 A1 = A + V * DT: REM  EQ. 4.11
1040 V1 = V - (G *  SIN (A1) / L) * DT: REM    EQ. 4.12
1050 XS = 50 + 150 * T
1060 YS = 96 - 20 * A
1070 & HPLOT XS,YS
1080 V = V1:A = A1
1090  NEXT T
2000  END
```

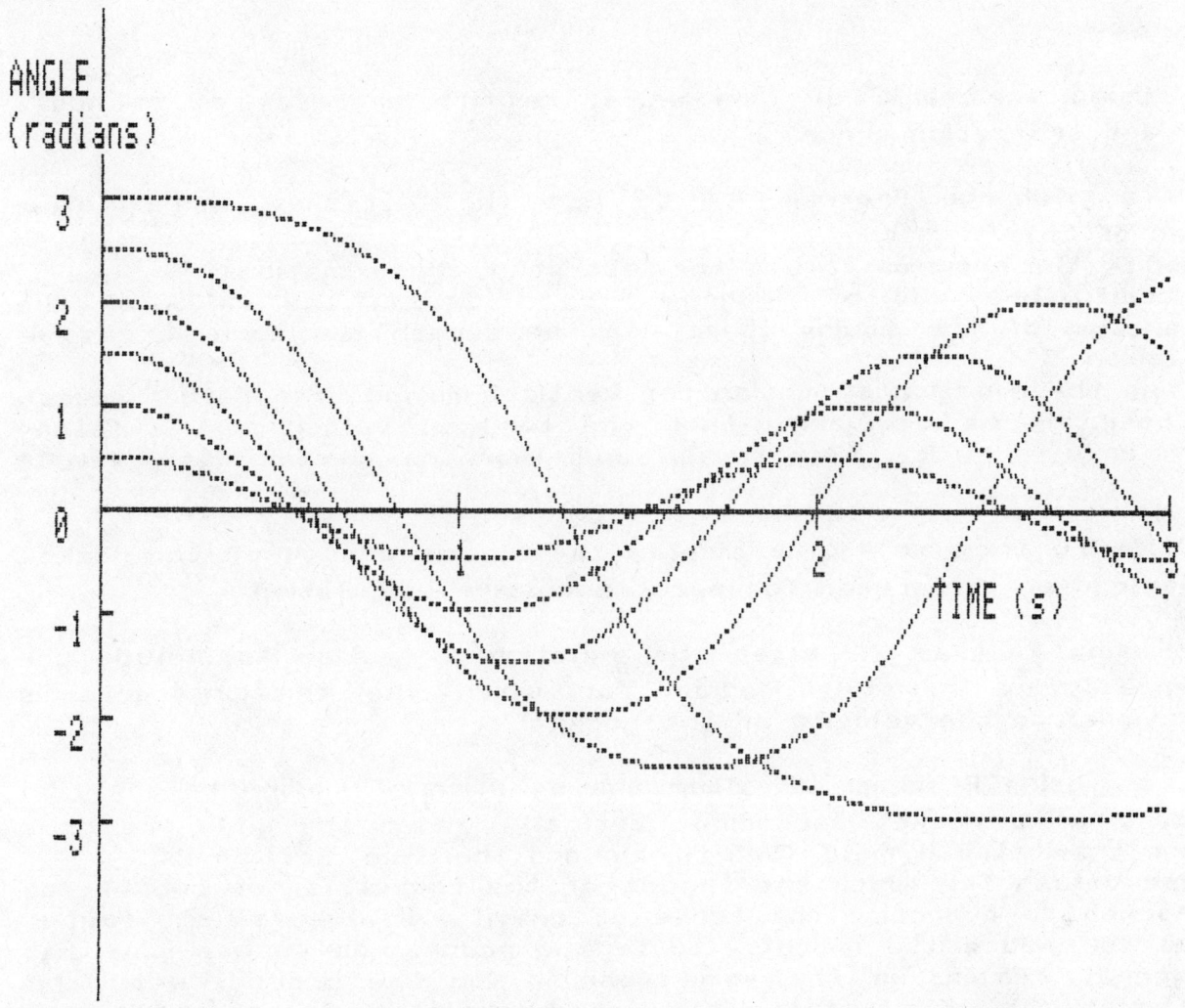

Figure 4.5. For this figure, Program 4.4 was modified so that the pendulum was started from rest with increasing values of initial angular position from 0.5 radians to 3.0 radians. The period of the motion increases when the pendulum swings with large amplitude. For small displacements, the period of the motion is independent of amplitude.

Exercises

4.1. Using the table of Chapter 4, identify examples of roundoff error and truncation error.

4.2. a. Change Program 4.1 to use the half-step approximation instead of Euler's method.
b. Modify the program to use the last-point approximation.
c. Modify line 1130 of Program 4.1 to display the value of the acceleration of the moving object on the screen for each successive calculation.
d. Using the equation of motion for vertical motion of a falling object in a resisting medium verify that the terminal velocity of a falling object is $v = -g/k$. (See Marion and Thornton, Section 2.4, Example 2.5.)

4.3. Modify Program 4.2 to display the kinetic energy of the rocket and payload on the screen for each successive calculation.

4.4. Using Program 4.2 alter the equation of motion to include the influence of a resisting medium for which the friction force is proportional to the velocity of the rocket.

4.5. a. Using Program 4.2 alter the equation of motion to apply to the case of a rocket ascending vertically under the influence of a uniform gravitational field. (See Marion and Thornton, Section 2.7.)
b. Use values for which the thrust of the rocket is not enough for the rocket to overcome the force of gravity until sufficient fuel is burned to reduce the weight. Write the program in such a way that the rocket remains in the same position (on the ground) until this condition is met. (Hint: if the calculated acceleration is negative, then the change in position is zero.)
c. Modify the program to continue to display the motion of the rocket after the fuel is exhausted. The rocket will then undergo free fall but with an initial upward velocity.

4.6. Modify Program 4.3 to plot graphs of both the velocity as a function of time and the distance traveled as a function of time as in Program 4.1.

4.7. Using Program 4.3 develop a technique to optimize the relative masses of the first and second stage in order to launch the payload with maximum velocity. Use the same design factors (exhaust velocity, rate of fuel consumption, mass of payload) as stated in the program but change the relative mass of the two stages while keeping the total mass constant.

4.8. Instead of graphing the angular position of the pendulum, modify Program 4.4 to produce a phase diagram of the motion of the pendulum.

4.9. a. Consider a pendulum subject to a friction force $-b\dot{\theta}$ acting on the pendulum bob. Using the last-point approximation, modify Program 4.4 to analyze the motion in this case.
b. Modify the program to produce a phase diagram for this system.

4.10. a. Modify Program 4.4 to graph the motion of a pendulum subject to a constant force parallel to the path of the pendulum.
b. Using the same initial conditions as those of Program 4.4 modify the program to allow the applied force to operate only until the pendulum passes through its equilibrium point ($\theta=0$).

4.11. a. Combine the principles of the simple pendulum and the equation of motion for a rocket to derive the equation of motion for a simple pendulum for which the pendulum bob is a rocket whose thrust is directed perpendicular to the pendulum suspension.
b. Modify Program 4.4 to plot a graph of the angular position of the pendulum as a function of time.

4.12. Write a program similar to Program 4.4 to analyze the motion of a nonlinear spring system. (See Marion and Thornton, Section 3.11, Example 3.5.)

4.13. Write a program similar to Program 4.4 to analyze the motion of a nonlinear oscillator in an asymmetric potential (See Marion and Thornton, Section 3.11.) such that the restoring force is: $F = -kx + k'x^3$

Problems

4.1. a. Write a program using the last point approximation to analyze the motion of a damped simple harmonic oscillator by graphing the displacement of the oscillator as a function of time. (See Marion and Thornton, Section 3.5 for the equation of motion of this system.) Using the program, demonstrate both underdamped and overdamped motion.
b. Modify the program to display the kinetic energy, potential energy and total energy of the system as the graph is drawn.
c. Graph the energy of the oscillator as a function of time.
d. Verify that energy loss of the system is equal to the work done by the friction force.
e. Modify the program to plot a phase diagram for this system.
f. Modify the program to include the effect of a sinusoidal driving force. (See Marion and Thornton, Section 3.6 for the equation of motion of this system.)
g. By eliminating the effect of the driving force after a time t has elapsed, investigate the response of the system to a sinusoidal impulse.

4.2. a. Write a program using the last point approximation to analyze the motion of a harmonic oscillator whose motion is affected by sliding friction.

b. Show graphically that the period of the oscillator is not affected by the friction and that the amplitude is decreased by the same amount in each cycle.
   HINT: The friction force is always in a direction opposite that of the velocity. Use the BASIC language statement -SGN(v) to assign the direction of the friction force.

4.3. For a simple harmonic oscillator, calculate the time averages of kinetic energy and potential energy. Verify that the quantities are equal when evaluated over a number of complete cycles of the oscillator.

4.4. Forces are often related to motion by empirical results rather than in terms of analytical equations. Figure A.2 shows the force on a projectile as a function of velocity. Using the top figure and the program created in Exercise A.14, write a program to plot a graph of velocity as a function of time for a projectile of this type falling under the influence of gravity. (See Marion and Thornton, Section 2.4, Example 2.5.) Depending on the value of velocity calculated in each time increment, use the appropriate value of force stored in an array [F(v/100)] when applying Euler's method (Eq. 3.8). For better results, interpolate between successive values of the force stored in the array in order to determine an accurate value of force for the calculated velocity. The weight of the falling projectile is 4.45 N (one pound). The terminal velocity occurs when the retarding force equals the weight of the falling projectile. Check your results by determining the terminal velocity using your program and comparing to the velocity on the graph for which the retarding force equals 4.45 N (one pound).

4.5. The point of support of a simple pendulum of length b moves on a massless rim of radius a that rotates with a constant angular velocity ω. The equation of motion of the pendulum is found to be:

$$\ddot{\theta} = (\omega^2 a/b)\cos(\theta - \omega t) - (g/b)\sin(\theta)$$

(See Marion and Thornton, Section 6.4, Example 6.4.) Write a program to plot the angular position θ as a function of time for this system.

4.6. A bead slides on a smooth wire bent in the shape of a parabola $z = cr^2$. The wire rotates about its vertical axis of symmetry with an angular velocity ω. The equation of motion of the bead is:

$$\ddot{r}(1 + 4c^2 r^2) + \dot{r}^2(4c^2 r) + r(2gc - \omega^2) = 0$$

(See Marion and Thornton, Section 6.4, Example 6.5.) Write a program to plot the vertical position of the bead z as a function of time.
b. Using your program, verify that the bead remains at a constant height z on the wire if $c = \omega^2/2g$.

CHAPTER 5

Motion in Two Dimensions

5.1 Introduction

For motion of a particle in two dimensions, applied forces can be resolved into components associated with each of the two coordinates used to describe the motion. As a result, two (not necessarily independent) equations of motion are required to determine the motion of the particle. These equations are usually written in terms of plane polar coordinates (r, θ) or cartesian coordinates (x, y) depending on the symmetry of the system being described.

5.2 Harmonic Oscillations in Two Dimensions

A simple harmonic oscillator can have two degrees of freedom with linear restoring forces directed along mutually perpendicular axes. In general, the force constants (k_x and k_y) along these directions need not be equal. The resulting motion is in the plane of the axes of the restoring forces with the motion of the oscillator described by "Lissajous curves." (See Marion and Thornton, Section 3.4.)

Equation 5.1 and Eq. 5.2 are the equations of motion for the two-dimensional undamped oscillator in cartesian coodinates.

$$d^2x/dt^2 = -k_x x/m \tag{5.1}$$

$$d^2y/dt^2 = -k_y y/m \tag{5.2}$$

5 · MOTION IN TWO DIMENSIONS

Applying Eq. 3.14 and Eq. 3.15 (half-step approximation) to Eq. 5.1 yields:

$$dx/dt\big)_{t+dt/2} = dx/dt\big)_{t-dt/2} - (k_x x\big)_t / m)dt \qquad (5.3)$$

$$x\big)_{t+dt} = x\big)_t + (dx/dt\big)_{t+dt/2})dt \qquad (5.4)$$

Equation 5.2 can be similarly rewritten in terms of the general form of the half-step approximation.

$$dy/dt\big)_{t+dt/2} = dy/dt\big)_{t-dt/2} - (k_y y\big)_t / m)dt \qquad (5.5)$$

$$y\big)_{t+dt} = y\big)_t + (dy/dt\big)_{t+dt/2})dt \qquad (5.6)$$

Program 5.1 uses Eq. 5.3, Eq. 5.4, Eq. 5.5, and Eq. 5.6 to evaluate the motion of a two-dimensional oscillator. The successive values of position (x, y) are scaled (line numbers 1080-1090) and plotted (line number 1100) to display the path of the motion. The resulting path yields a Lissajous curve which can be modified by using different sets of initial conditions or by varying the relative values of the force constants k_x and k_y.

```
90  REM   PROGRAM (5.1)
100 REM      ***** SET UP GRAPHICS CHARACTERISTICS *****

110 & HGR2 : & HCOLOR= 15: & B COLOR= 0: & PRINT : & MODE(1)
300 REM
                ***** SET UP SCREEN DISPLAY *****

500 REM
                ***** SPECIFY INITIAL CONDITIONS *****

510 V1 = .5: REM   INITIAL X COMPONENT OF VELOCITY
520 V3 = 0: REM    INITIAL Y COMPONENT OF VELOCITY
530 X = .7: REM    INITIAL POSITION RELATIVE TO HORIZONTAL (X) AXIS
540 Y = .7: REM    INITAL POSITION RELATIVE TO VERTICAL (Y) AXIS
550 KX = 9: REM    SPRING CONSTANT IN X DIRECTION
560 KY = 16: REM   SPRING CONSTANT IN Y DIRECTION
570 M = 1: REM     MASS OF OSCILLATOR
580 DT = .01: REM  TIME INCREMENT
1000 REM
                ***** CALCULATE AND PLOT POSITION *****

1010 V1 = V1 - (KX / M) * X * DT / 2: REM  Eq. 5.3
1020 V3 = V3 - (KY / M) * Y * DT / 2: REM  Eq. 5.5
1030 FOR T = 0 TO 100 STEP DT
1040 X1 = X + V1 * DT: REM  Eq. 5.4
1050 Y1 = Y + V3 * DT: REM  Eq. 5.6
```

5.2 HARMONIC OSCILLATIONS IN TWO DIMENSIONS

```
1060 V2 = V1 - (KX / M) * X1 * DT
1070 V4 = V3 - (KY / M) * Y1 * DT
1080 XS = 280 + 160 * X
1090 YS = 96 - 80 * Y
1100 &   HPLOT XS,YS
1110 V1 = V2:V3 = V4
1120 X = X1:Y = Y1
1130  NEXT T
2000  END
```

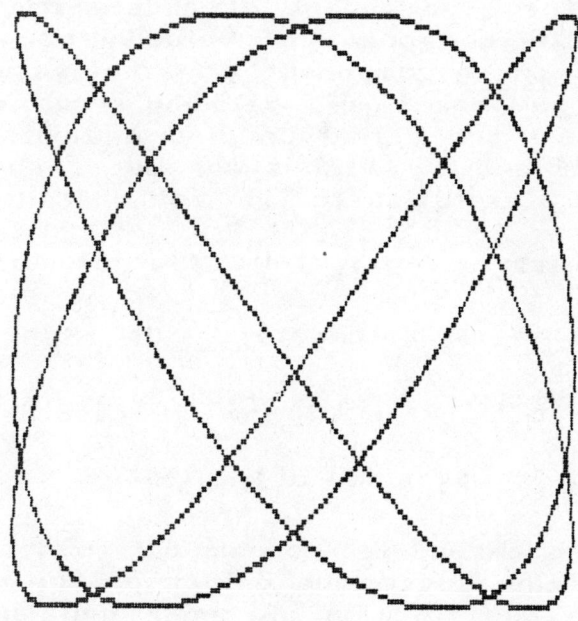

Figure 5.1. This Lissajous curve can be altered by changing the initial conditions of the motion or the mechanical description of the system in Program 5.1. For example, if the values of the spring constants k_x and k_y are equal, the resulting figure is an ellipse. The eccentricity and orientation of the ellipse are determined by the initial location and velocity of the particle.

5.3 Foucault Pendulum

The action of the Coriolis effect on the motion of a pendulum causes precession of the plane of oscillation of the pendulum. Equation 5.7 and Eq. 5.8 are the equations of motion for a pendulum which has horizontal (x and y) components of motion which are small compared to the vertical (z) height of the pendulum suspension. (See Marion and Thornton, Section 9.4, Example 9.5.) For a pendulum on a rotating spherical surface (such as the earth), ω is the angular velocity of the rotation and θ is the latitude at which the pendulum is suspended.

$$d^2x/dt^2 = -gx/l + 2\dot{y}\omega \sin(\theta) \qquad (5.7)$$

$$d^2y/dt^2 = -gy/l - 2\dot{x}\omega \sin(\theta) \qquad (5.8)$$

The equation for the x-component of acceleration (Eq. 5.7) contains a term involving $\dot{y}$, the component of velocity parallel to the y axis and the equation for the y-component of acceleration contains a term involving $\dot{x}$, the component of velocity parallel to the x-axis. Equations of this type are called coupled equations. The last point approximation (Eq. 3.16 and Eq. 3.17) is applied as shown below in order to generate a numerical solution to this set of coupled equations.

$$dx/dt\}_{t+dt} = dx/dt\}_t + (-gx\}_t/l + 2[(dy/dt)\}_t \omega \sin(\theta)])dt \qquad (5.9)$$

$$x\}_{t+dt} = x\}_t + (dx/dt\}_{t+dt})dt \qquad (5.10)$$

$$dy/dt\}_{t+dt} = dy/dt\}_t + (-gx\}_t/l - 2[(dx/dt)\}_{t+dt} \omega \sin(\theta)])dt \qquad (5.11)$$

$$y\}_{t+dt} = y\}_t + (dy/dt\}_{t+dt})dt \qquad (5.12)$$

These equations are applied in Program 5.2 to determine the path of the pendulum by plotting successive values of the position (x, y). The graph resulting from the solution of these equations represents the projection of the pendulum motion onto the horizontal (x-y) plane. The vertical (z) component of the motion is negligible in this case.

```
90  REM   PROGRAM(5.2)
100 REM      ***** SET UP GRAPHICS CHARACTERISTICS *****

110 &  HGR2 : &  HCOLOR= 15: &  B COLOR= 0: &   PRINT : & MODE(1)
300 REM
                  *****  SET UP SCREEN DISPLAY *****


500 REM
                 ***** SPECIFY INITIAL CONDITIONS *****


510 X = 1
```

5.3 CENTRAL FORCE MOTION (POLAR COORDINATES)

```
520 Y = 0
530 V1 = 0
540 V3 = 0
550 L = 10
560 W = 7.27 * 10 ^ - 2: REM   ANGULAR VELOCITY OF ROTATING SYSTEM
(1000 TIMES THE ANGULAR VELOCUTY OF THE EARTH)
570 G = 9.8
580 DT = .05
1000  REM
                ***** CALCULATE AND PLOT VELOCITY AND POSITION *****

1010  FOR T = 0 TO 1000 STEP DT
1020  V2 = V1 + ( - G * X / L + 2 * W * V3) * DT
1030  V4 = V3 + ( - G * Y / L - 2 * W * V2) * DT
1040  X1 = X + V2 * DT
1050  Y1 = Y + V4 * DT
1060  XS = 280 + 160 * X
1070  YS = 96 - 80 * Y
1080   &  HPLOT XS,YS
1090  V1 = V2:V3 = V4
1100  X = X1:Y = Y1
1110  NEXT T
2000  END
```

5.4 Central Force Motion
   (Two-Body Motion in Polar Coordinates)

Because of the spherical symmetry of the system, the equations of motion of a two-body system moving under the influence of a central force F(r) are expressed most clearly when written in terms of polar coordinates (r, θ).

$$l = mr^2\dot{\theta} = \text{const} \qquad (5.13)$$

$$m(\ddot{r} - r\dot{\theta}^2) = F(r) \qquad (5.14)$$

In these equations, m is the reduced mass of the two-particle system, $\dot{\theta}$ represents the angular velocity of the system and $\ddot{r}$ represents the radial acceleration along the line directed between the interacting particles. Equation 5.13 defines the angular momentum (l) which is a constant for the system. (See Marion and Thornton, Section 7.3 and Section 7.4.)

An expression for the angular velocity $\dot{\theta}$ of the system can be derived from Eq. 5.13.

$$\dot{\theta} = l/mr^2 \qquad (5.15)$$

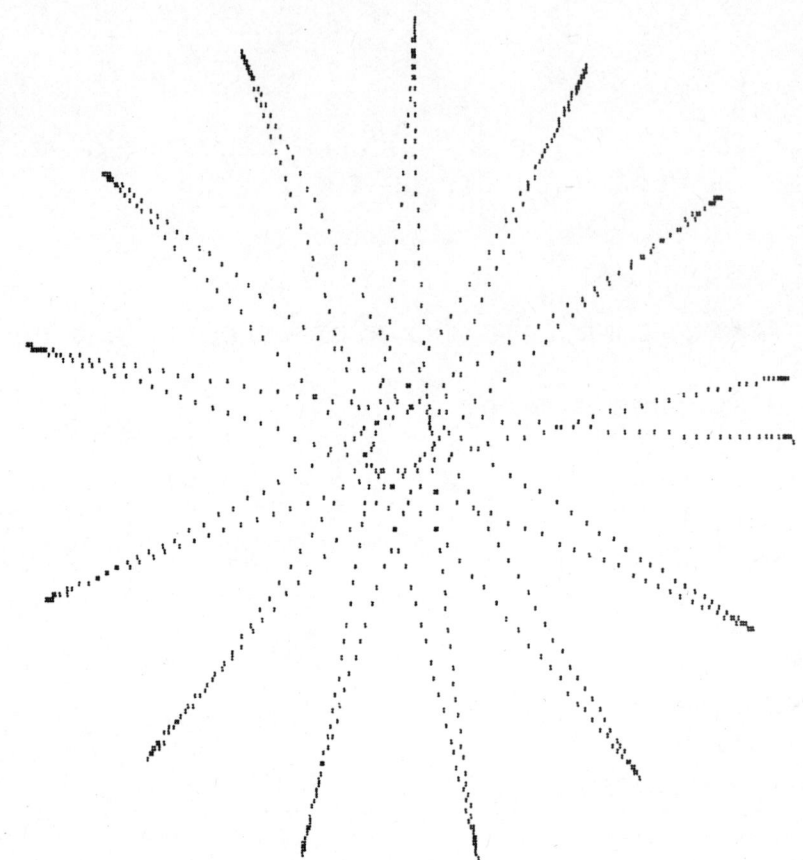

Figure 5.2. This graph represents the motion of a Foucault pendulum projected onto the horizontal plane. The direction of ω, the angular velocity of the rotating system, is upward, out of the page (along the z axis). The magnitude of ω used to produce this figure is 1000 times the angular velocity of the earth. Thus the resulting rate of precession is 1000 times greater than the precession rate of a Foucault pendulum on earth.

Substitution of this expression into Eq. 5.14 yields:

$$\ddot{r} = F(r)/m + l^2/mr^3 \qquad (5.16)$$

Program 5.3 evaluates the equations of motion (Eq. 5.15 and Eq. 5.16) using the last point approximation (Eq. 3.16 and Eq. 3.17). In order to plot the paths of satellites of the earth, the y axis is scaled with one screen pixel representing a distance of 2 million meters. The central force acting between the particles has the familiar form of Newton's Law of Gravitation: $F = -GMm/r^2$.

The value of the angular momentum l of the system is calculated in line number 570 using the initial conditions of the motion. Equation 5.16 is evaluated in line number 1025 to determine the radial velocity V.

5.4 CENTRAL FORCE MOTION (POLAR COORDINATES)

The distance of separation R is determined in line number 1030. Equation 5.15 is then evaluated to determine the angular position A in line number 1040. The cartesian coordinates (X,Y) corresponding to the calculated values of R and A are determined in line number 1060. The familiar scaling and positioning of the figure is performed in line number 1040.

```
90  REM   PROGRAM (5.3)
100  REM     ***** SET UP GRAPHICS CHARACTERISTICS *****
110  & HGR2 : & HCOLOR= 15: & B COLOR= 0: & PRINT : & MODE(1)
300  REM
                 *****   SET UP SCREEN DISPLAY *****

310  & HPLOT 1,96 TO 550,96
320  & HPLOT 280,1 TO 280,191
330  FOR X = 10 ^ 7 TO 6 * 10 ^ 7 STEP 10 ^ 7
340  XS = 280 + X * 4 * 10 ^  - 6: REM   SCALE HORIZONTAL AXIS
350  & GOTO XS,96: & HPLOT XS,96 - 2 TO XS,96 + 2: & GOTO XS - 6,96 + 4: PRINT X / 10 ^ 7
355  NEXT X
357  FOR Y = 10 ^ 7 TO 4 * 10 ^ 7 STEP 10 ^ 7
360  YS = 96 - Y * 2 * 10 ^  - 6: REM   SCALE VERTICAL AXIS
370  & GOTO 280,YS: & HPLOT 280 - 4,YS TO 280 + 4,YS: & GOTO 280 + 6,YS - 2: PRINT Y / 10 ^ 7
380  NEXT Y
390  & GOTO 390,108: PRINT "R(10^7m)"
410  FOR A = 0 TO 6.28 STEP .04:XC = 280 + 26 * COS (A):YC = 96 - 13 * SIN (A): & HPLOT XC,YC: NEXT A: REM   DRAW CIRCLE AT CENTER OF SCREEN
500  REM
                 ***** SPECIFY INITIAL CONDITIONS *****

510  R = 5.23 * 10 ^ 7: REM   INITIAL VALUE OF R (DISTANCE FROM EARTH TO SATELLITE)
515  A = 3: REM INITIAL ANGULAR POSITION
520  V = 0: REM   INITIAL RADIAL VELOCITY
530  VA = 3.5 * 10 ^  - 5: REM         INITIAL ANGULAR VELOCITY
540  G = 6.67 * 10 ^  - 11: REM   UNIVERSAL GRAVITATIONAL CONSTANT
550  M = 6 * 10 ^ 24: REM   MASS OF EARTH
560  M1 = 100: REM    MASS OF SATELLITE (REDUCED MASS OF SYSTEM)
570  L = M1 * (R ^ 2) * VA: REM    ANGULAR MOMENTUM OF SYSTEM
580  DT = 500: REM    TIME INCREMENT
1000  REM
              ***** CALCULATE AND PLOT VELOCITY AND POSITION *****

1020  FOR T = 0 TO 800000 STEP DT
```

```
1025 V1 = V + (((  - G * M) / (R ^ 2)) + ((L ^ 2) / ((M1 ^ 2) * (R ^
3)))) * DT
1030 R1 = R + V1 * DT
1040 A1 = A + (L / (M1 * (R1 ^ 2))) * DT
1060 X = R *  COS (A):Y = R *  SIN (A)
1070 XS = 280 + X * 4 * (10 ^  - 6):YS = 96 - Y * 2 * (10 ^  - 6)
1080 &   HPLOT XS,YS
1090 V = V1:R = R1:A = A1
1100  NEXT T
2000  END
```

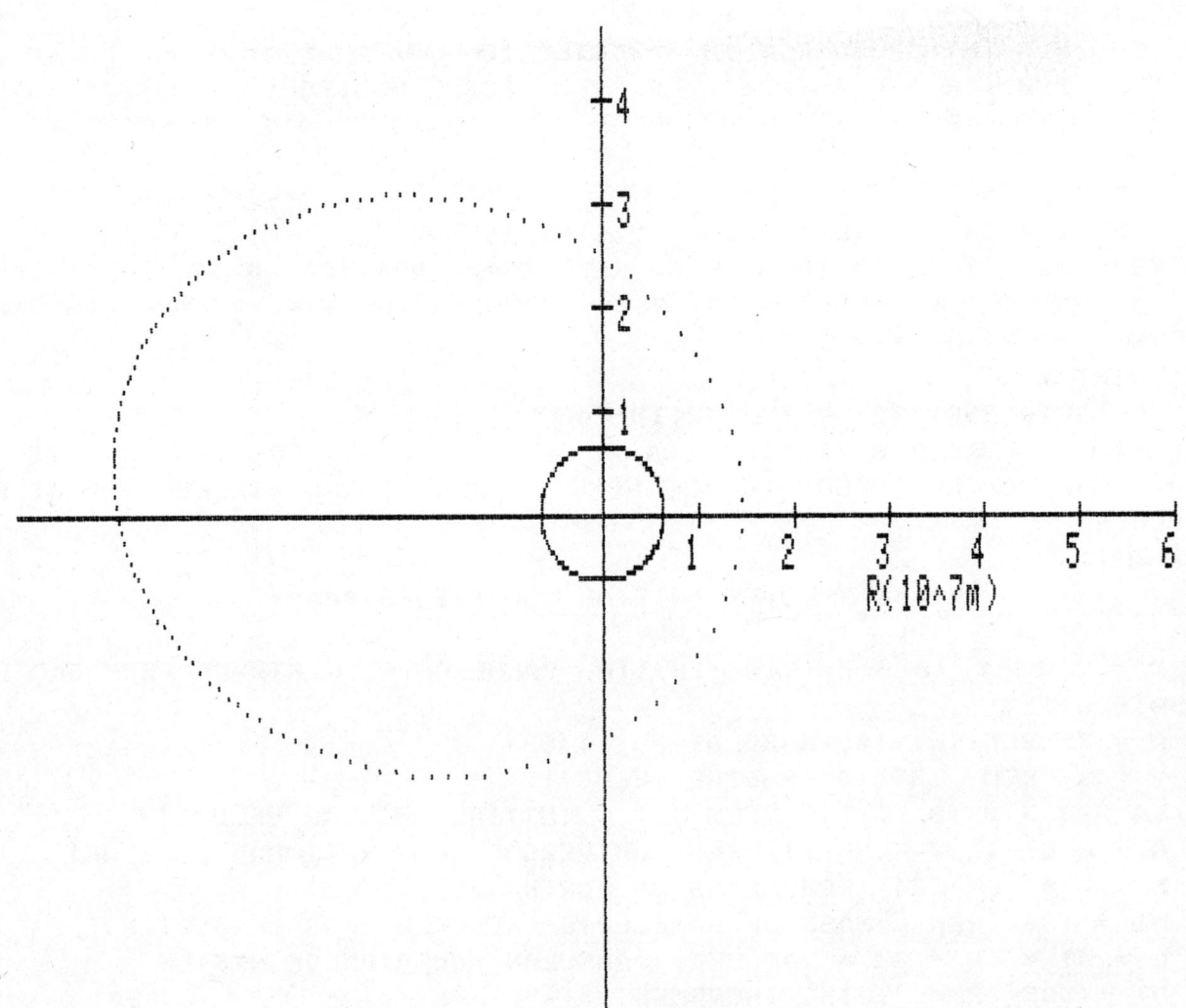

Figure 5.3. Orbital path of a satellite of the earth with initial conditions of Program 5.3.

5.5 Orbital Boost

If a satellite is aimed so that it travels "behind" a massive object moving in a straight line, the satellite gains energy as a result of the interaction. (See Marion and Thornton, Section 7.11.) This "flyby" or "slingshot" effect is being used to propel the spacecraft Voyager 1 and Voyager 2 into interstellar space after they pass near several of the outer planets. Voyager 1 is already on its way to interstellar space after passing near Jupiter and Saturn. The path of Voyager 2 was altered first by an encounter with Jupiter, then Saturn followed by Uranus. By carefully controlling these encounters, this spacecraft was directed toward Neptune which it will encounter in August, 1989 before it also passes into interstellar space. In these cases, the gravitational boost added enough energy to the motion of the spacecraft to allow them to exceed the escape velocity of our solar system.

Program 5.4 illustrates the motion of a satellite with a mass of 100 kg encountering the earth's moon. Because the mass of the moon is several orders of magnitude geater than the mass of the satellite in this example, the moon is considered to move in a straight line at constant velocity. The satellite is initially at rest (as seen from the earth) as the moon moves forward with a velocity equal to its orbital velocity relative to the earth. The program plots both the path of the moon (a straight line) and the path of the accelerated satellite by evaluating the effect of the gravitational force on the path of the satellite. The figure on the computer screen thus represents the path seen by an observer at rest with respect to the earth.

In Program 5.4, the path of the moon is directed along the x axis of the cartesian coordinate system used to describe the motion. The variable A represents the position of the moon on the x axis. The satellite is located at the point (X, Y). The Pythagorean theorem is applied in line number 1030 to calculate the distance of separation (R) between the two objects in terms of their relative locations. The magnitude of the gravitational force (F) is then calculated in line number 1040. The horizontal component of force affecting the motion of the satellite (FX) is calculated in line number 1050 by multiplying the gravitational force (F) by [(A-X)/R]; the ratio of the horizontal component of the distance of separation to the total distance of separation. (This ratio is the cosine of the angle between the line separating the two objects and the horizontal axis.) The vertically directed force component (FY) is similarly calculated in line number 1060. The last point approximation is then used in line numbers 1070-1100 to evaluate the equations of motion and to determine the horizontal and vertical components of the motion of the satellite.

```
90   REM   PROGRAM (5.4)
100  REM      ***** SET UP GRAPHICS CHARACTERISTICS *****

110  & HGR2 : & HCOLOR= 15: & B COLOR= 0: & PRINT : & MODE(1)
300  REM
                ***** SET UP SCREEN DISPLAY *****
```

58   5 · MOTION IN TWO DIMENSIONS

```
310   & HPLOT 280,1 TO 280,190
320   FOR X = - 9 TO 9
330   XS = 280 + X * 90 / 3: REM      SCALE SCREEN HORIZONTALLY(30
PIXELS=1*10^6M)
340   & HPLOT XS,96 + 3 TO XS,96 - 3: & GOTO XS,96 + 5: PRINT X
350   NEXT X
360   & GOTO 300,110: PRINT "DISTANCE (10^6m)"
500   REM
               ***** SPECIFY INITIAL CONDITIONS *****

510  MM = 7.4 * 10 ^ 22: REM    MASS OF THE MOON (KG)
520  M = 100: REM  MASS OF SATELLITE (KG)
530  G = 6.67 * 10 ^ - 11: REM    UNIVERSAL GRAVITATIONAL CONSTANT
540  VA = 930: REM      VELOCITY OF MOON (HORIZONTAL MOTION RELATIVE
TO THE SCREEN)
545  VB = 0
550  VX = 0: REM       INITIAL HORIZONTAL COMPONENT OF VELOCITY OF
SATELLITE (RELATIVE TO SCREEN)
560  VY = 0: REM        INITIAL VERTICAL COMPONENT OF VELOCITY OF
SATELLITE (RELATIVE TO SCREEN)
570  A = - 9 * 10 ^ 6: REM   INITIAL X COORDINATE OF MOON
580  B = 0: REM   INITIAL Y COORDINATE OF MOON
590  X = 0: REM     INITIAL X COORDINATE OF SATELLITE
600  Y = - 3 * 10 ^ 6: REM    INITIAL Y COORDINATE OF SATELLITE
610  DT = 100: REM  TIME INCREMENT
1000  REM
           *****  CALCULATE AND PLOT VELOCITY AND POSITION *****

1010  FOR T = 0 TO 12000 STEP DT
1020  A = A + VA * DT:UM = 280 + A * (90 / (3 * 10 ^ 6)): & HPLOT
UM,96: REM        PLOT PATH OF MOON
1030  R = SQR ((A - X) ^ 2 + (B - Y) ^ 2): REM    DISTANCE BETWEEN
MOON AND SATELLITE (M)
1040  F = G * MM * M / (R * R): REM    GRAVITATIONAL FORCE BETWEEN
MOON AND SATELLITE
1050  FX = F * (A - X) / R: REM     X-COMPONENT OF GRAVITATIONAL FORCE
1060  FY = F * (B - Y) / R: REM     Y-COMPONENT OF GRAVITATIONAL FORCE
1070  VX = VX + (FX / M) * DT
1080  VY = VY + (FY / M) * DT
1090  X1 = X + VX * DT: REM  LAST POINT APPROXIMATION
1100  Y1 = Y + VY * DT
1110  XS = 280 + X * (90 / (3 * 10 ^ 6))
1120  YS = 96 - Y * (45 / (3 * 10 ^ 6))
1130  &  HPLOT XS,YS
```

```
1140  X = X1:Y = Y1
1150   NEXT T
2000   END
```

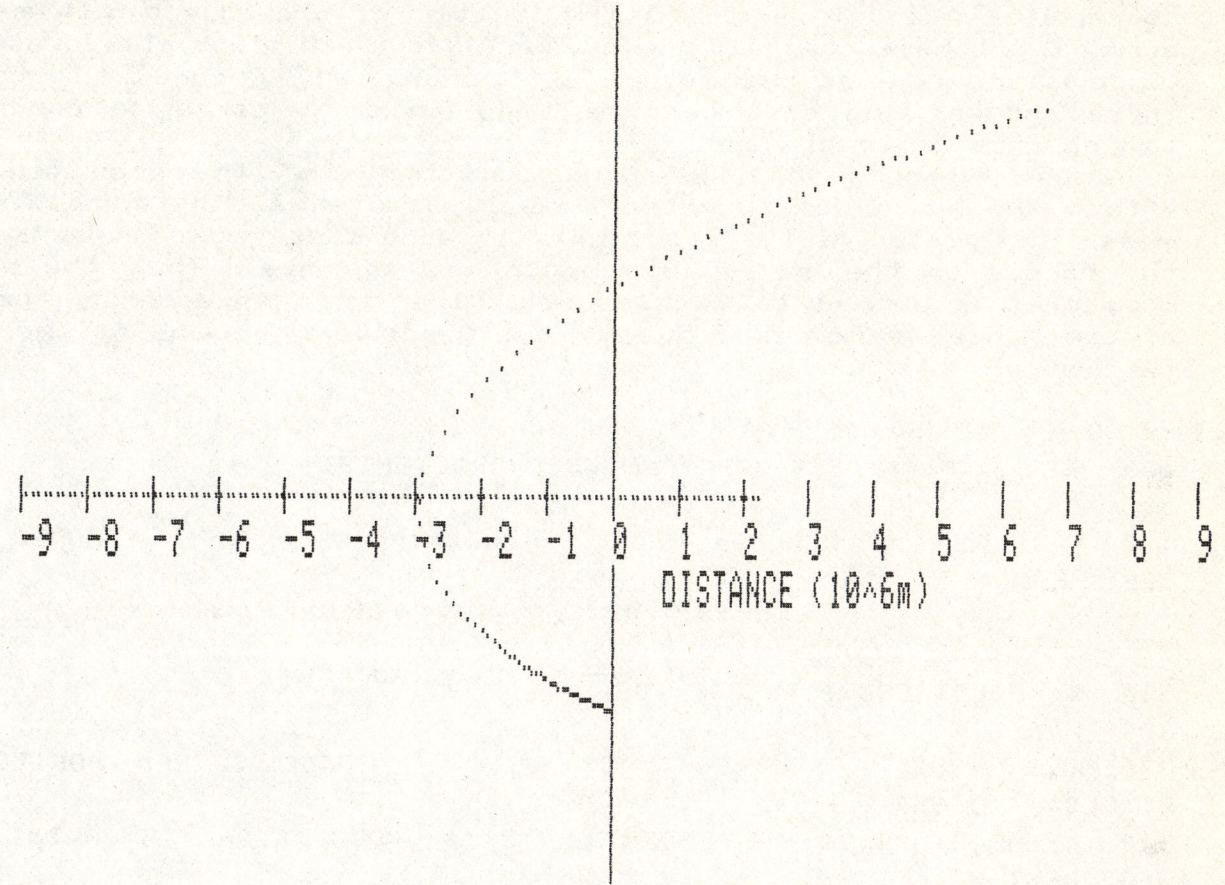

Figure 5.4. Relative to the computer screen, the satellite is originally at rest on the negative y-axis and gains kinetic energy as it interacts gravitationally with the moon. After beginning at a distance of 9 million meters from the y axis at t=0, the path of the moon continues at a constant velocity in the positive direction along the x-axis. With different sets of initial conditions, the satellite can experience either energy loss or energy gain. Energy loss occurs if the initial conditions of the motion are modified so that the satellite moves in front of the moon.

5.6 Central Force Motion (Cartesian Coordinates)

Program 5.4 describes one-body motion. The deviation of the path of the heavy object (moon) is negligible compared to the motion of the lighter object (satellite). The equation of motion was applied only to the motion of the lighter object. In order to calculate the influence of the central force on the path of both objects, the equation of motion must be applied to both objects in accordance with Newton's Third Law. (See Marion and Thornton, Section 2.2.)

5 · MOTION IN TWO DIMENSIONS

To modify Program 5.4 in order to correctly apply Newton's Third Law, the components of the gravitational force must affect the motion of both objects. This change in the program can be made by altering line numbers 1070 through 1140. In these program lines, X and Y continue to represent the coordinates of the the object with mass M (satellite). The coordinates A and B represent the location of the object with mass MM. In addition to these changes, line number 1020 which plots the location of the (undeviated) path of the moon, must be deleted. Program 5.5 incorporates these changes and describes two-body motion in cartesian coodinates for two objects of equal mass.

In line numbers 1070-1080, the components of the gravitational force affect the motion of both the mass M (located at the point X,Y) and the mass MM (located at the point A,B). In accordance with Newton's Third Law the forces on the masses are equal and opposite. Thus the same force component is applied to each object. The sign (representing the direction of the force) is reversed due to the opposite directions of the forces on the two objects.

```
90   REM   PROGRAM (5.5)
100  REM        ***** SET UP GRAPHICS CHARACTERISTICS *****

110  & HGR2 : & HCOLOR= 15: & B COLOR= 0: &  PRINT : & MODE(1)
300  REM
                    ***** SET UP SCREEN DISPLAY *****

310  & HPLOT 280,1 TO 280,190
320  FOR X = - 9 TO 9
330  XS = 280 + X * 90 / 3: REM         SCALE SCREEN HORIZONTALLY(30
PIXELS=1*10^6M)
340  & HPLOT XS,96 + 3 TO XS,96 - 3: &  GOTO XS,96 + 5: PRINT X
350  NEXT X
360  & GOTO 300,110: PRINT "DISTANCE (10^6m)"
500  REM
                    ***** SPECIFY INITIAL CONDITIONS *****

510 MM = 7.4 * 10 ^ 22: REM    MASS OF THE MOON (KG)
520 M = 7.4 * 10 ^ 22: REM     MASS OF SATELLITE (KG)
530 G = 6.67 * 10 ^ - 11: REM    UNIVERSAL GRAVITATIONAL CONSTANT
540 VA = 930: REM   VELOCITY OF MOON (HORIZONTAL MOTION RELATIVE TO THE
SCREEN)
545 VB = 0
550 VX = 0: REM    INITIAL HORIZONTAL COMPONENT OF VELOCITY OF SATELLITE
(RELATIVE TO SCREEN)
560 VY = 0: REM      INITIAL VERTICAL COMPONENT OF VELOCITY OF SATELLITE
(RELATIVE TO SCREEN)
570 A = - 9 * 10 ^ 6: REM         INITIAL X COORDINATE OF MOON
580 B = 0: REM       INITIAL Y COORDINATE OF MOON
590 X = 0: REM    INITIAL X COORDINATE OF SATELLITE
```

5.6 CENTRAL FORCE MOTION (CARTESIAN COORDINATES)

```
600 Y =  - 3 * 10 ^ 6: REM     INITIAL Y COORDINATE OF SATELLITE
610 DT = 50: REM                TIME INCREMENT
1000  REM
           *****   CALCULATE AND PLOT VELOCITY AND POSITION *****

1010   FOR T = 0 TO 15000 STEP DT
1030 R =   SQR ((A - X) ^ 2 + (B - Y) ^ 2): REM    DISTANCE BETWEEN MOON
AND SATELLITE (M)
1040 F = G * MM * M / (R * R): REM    GRAVITATIONAL FORCE BETWEEN MOON
AND SATELLITE
1050 FX = F * (A - X) / R: REM    X-COMPONENT OF GRAVITATIONAL FORCE
1060 FY = F * (B - Y) / R: REM    Y-COMPONENT OF GRAVITATIONAL FORCE
1070 VX = VX + (FX / M) * DT:VA = VA - (FX / MM) * DT
1080 VY = VY + (FY / M) * DT:VB = VB - (FY / MM) * DT
1090 X1 = X + VX * DT:A1 = A + VA * DT
1100 Y1 = Y + VY * DT:B1 = B + VB * DT
1110 XS = 280 + X * (90 / (3 * 10 ^ 6)):AS = 280 + A * (90 / (3 * 10 ^ 6))
1120 YS = 96 - Y * (45 / (3 * 10 ^ 6)):BS = 96 - B * (45 / (3 * 10 ^ 6))
1130   &   HPLOT XS,YS: &   HPLOT AS,BS
1140 X = X1:Y = Y1:A = A1:B = B1
1150   NEXT T
2000   END
```

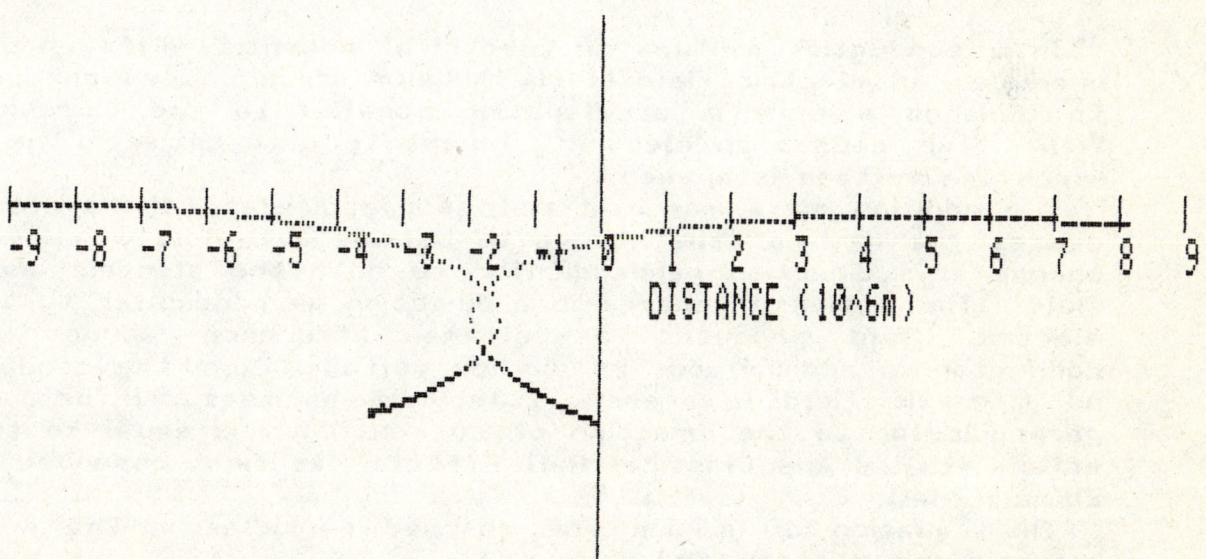

Figure 5.5. The motion of two objects of equal mass is shown. The initial conditions of the motion are exactly the same as those for the one-body motion shown in Figure 5.4. However, both M and MM, the masses of the moving objects, are 7.4*10^22 kg (the mass of the earth's moon) in this example.

5.7 Three-Body Motion (Two Dimensions)

The addition of a third movable object to a system of objects interacting through a central force results in a system for which the equations of motion have never been successfully solved in closed analytical form by elementary means after 200 years of effort by a most distinguished list of physicists.

A simple extension of the procedures developed in the preceding programs provides a technique for developing a numerical solution to this infamous problem. A program which performs this calculation is not presented here in order that students can develop the solution for themselves.

The outline of the procedure to extend the previous program to account for the motion of a third object is presented below.

    a. Calculate the relative distances between each pair of objects (see line number 1030).

    b. Calculate the forces between each pair of interacting objects (see line number 1040)

    c. Calculate the components of each force (see line numbers 1050-1060)

    d. Find the total of the force components applied to each of the three interacting objects.

    e. Evaluate the equation of motion for each object. (see line numbers 1070-1100)

    f. Scale the sceen and use graphics to plot the position of each object. (see line numbers 1110-1130)

5.8 Classical Hall Effect

In a conducting medium, an electrical potential difference (voltage) produces an electric field in the medium which causes charge carriers to undergo a uniform acceleration parallel to the direction of the field. This causes an electric current to flow between the points at which the voltage is applied.

The addition of a magnetic field perpendicular to the electric field causes the path of the charge to become a cycloid which carries the charge in a direction perpendicular to both the electric and magnetic field. The motion of charge in a direction perpendicular to the applied electric field produces a voltage difference across the solid conductor in addition to the applied voltage along the conductor. (If no magnetic field is present, the voltage measured between points perpendicular to the direction of current flow is equal to zero.) This effect (called the Classical Hall Effect) was first observed in 1879 by Edwin H. Hall.

The equation of motion for charged particles in the presence of electric and magnetic fields is:

$$m d^2 r/dt^2 = q(E + v \times B) \qquad (5.15)$$

If the electric field vector (E) is directed parallel to the x axis and the magnetic field vector (B) is directed parallel to the z axis, Eq. 5.15 becomes:

5.8 CLASSICAL HALL EFFECT

$$md^2x/dt^2 = q(E_x + v_y B_z) \qquad (5.16)$$

$$md^2y/dt^2 = q(v_x B_z) \qquad (5.17)$$

$$md^2z/dt^2 = 0 \qquad (5.18)$$

Program 5.6 uses a variation of the mid-point approximation to evaluate Eq. 5.16 and Eq. 5.17 and plot the motion of an electron in the presence of the electric field E and magnetic field B described above. The electric field directed parallel to the x-axis is parallel to the screen and the magnetic field directed parallel to the z-axis points out of the screen. Equation 5.18 is not used to evaluate the z-component of the motion. This component of velocity is constant and does not affect the projection of the motion plotted in the x-y plane (the plane of the screen).

The path of the electron is a prolate cycloid (trochoid) whose guiding center moves the electron parallel to the y-axis (toward the top of the screen). If the region depicted on the screen were a solid conductor, the accumulation of electrons would make the voltage in the upper half of the screen negative relative to the voltage in the lower half of the screen where the number of electrons is depleted.

```
90  REM   PROGRAM (5.6)
100 REM         ***** SET UP GRAPHICS CHARACTERISTICS *****
110  & HGR2 : & HCOLOR= 15: & B COLOR= 0: &  PRINT : & MODE(1)
300 REM            *****  SET UP SCREEN DISPLAY *****
310  &   HPLOT 380,20 TO 430,20: &   GOTO 430,17: PRINT ">    (+X
direction)": &  GOTO 360,17: PRINT "E"
320  &  GOTO 360,30: PRINT "B": &  GOTO 380,27: PRINT ".": &  GOTO
450,29: PRINT "(+Z direction)"
330  FOR A =  - 2 TO 2
335  &  HPLOT 1,96 TO 550,96
340 X = 280 + A * 100
350  &  HPLOT X,96 + 3 TO X,96 - 3: &  GOTO X,96 + 5: PRINT A
360  NEXT A
370  &  HPLOT 280,8 TO 280,190: &  GOTO 278,1: PRINT "Y"
380  &  GOTO 400,96 + 10: PRINT "X (10^-4m)"
500 Q = 1.6 * (10 ^  - 19)
510 M = 9.1 * (10 ^  - 31)
520 E = 300 : B = .01
540 V1 = 70000
550 V3 = 70000
560 DT = 1 * (10 ^  - 10)
570 X = 0:Y = 0
1000 REM    ***** CALCULATE AND PLOT VELOCITY AND POSITION *****
1010 V1 = V1 + (Q / M) * (E - V3 * B) * DT / 2
1020 V3 = V3 + (Q / M) * (V1 * B) * DT / 2
```

```
1030  FOR T = 0 TO 1.1 * 10 ^ - 8 STEP DT
1040  X1 = X + V1 * DT
1050  Y1 = Y + V3 * DT
1060  V2 = V1 + (Q / M) * ((E) - (V3 * B)) * DT
1070  V4 = V3 + (Q / M) * (V2 * B) * DT
1080  XS = 280 + X * (10 ^ 6)
1090  YS = 180 - Y * (.5 * 10 ^ 6)
1100  &   HPLOT XS,YS
1110  X = X1:Y = Y1:V1 = V2:V3 = V4
1120  NEXT T
2000  END
```

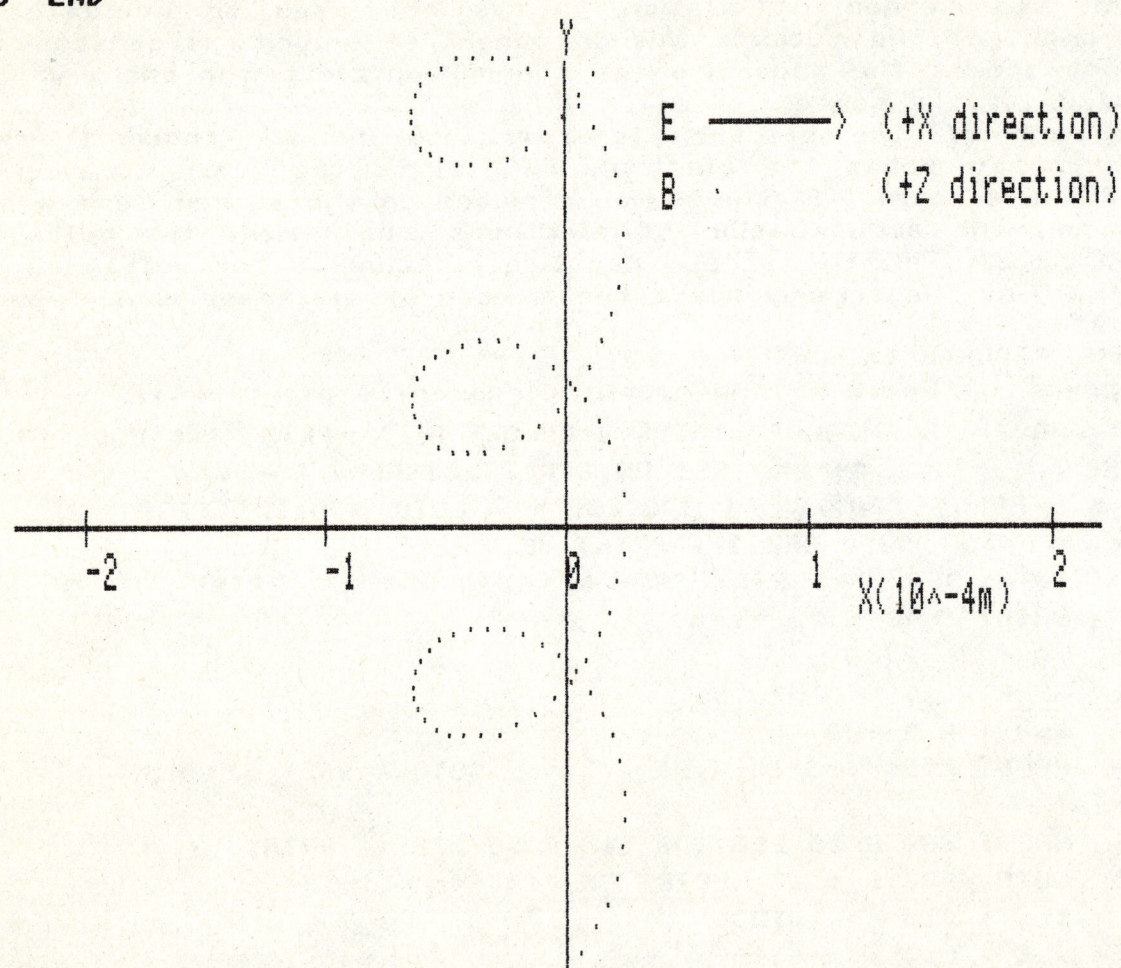

Figure 5.6. The dotted line represents the path of an electron moving under the influence of an electric field E and a magnetic field B. The magnetic field strength is .01 T directed out of the page (+z direction). The electric field strength is 300 N/C directed parallel to the page (+x direction) in this example. The initial position of the electron was on the y-axis at the bottom of the figure. The initial velocity of the electron had equal components in the +x direction and in the +y direction.

Exercises

5.1. a. Show that the path of the Foucault pendulum shown in Figure 5.2 can take the shape of a "figure-8" when the pendulum is started with a small component of velocity perpendicular to the plane of swing of the pendulum.
b. Show that if the component of velocity perpendicular to the plane of swing is large enough, the pendulum will not swing but will instead move in a circular path.
c. Modify Program 5.2 to analyze the motion of a pendulum with a "Y-shaped" suspension by making the value of the length of the pendulum (l) different in each of the coupled equations (Eq. 5.7 and Eq. 5.8).

5.2. a. Using Program 5.3, verify that a circular orbit is produced for initial conditions such that $r = GM/v^2$.
b. A small mass is released from rest at a distance from the earth equal to the orbital radius of the moon. What amount of time must elapse for the object to fall a distance of 108 meters toward the earth? What is the speed of the object at that point? Verify that this speed is consistent with the prediction of the law of conservation of energy.

5.3. Modify Program 5.3 to analyze the motion of a two-body system under the influence of a force: $F = -GMm/r^3$.

5.4. a. Modify Program 5.3 to add the influence of a small force, $F = -kr$, in addition to the gravitational force. The orbits drawn in this case show the effect of orbital motion inside a dust cloud (ignoring friction). At one time, this effect was thought to be responsible for the precession of the orbit of Mercury. (See Marion and Thornton, Section 7.9.)

5.5. a. Using Program 5.3, show that the total energy of the orbiting particle is constant by direct calculation and display of the total energy with each successive calculation.
b. Verify the virial theorem by direct calculation of the average values of kinetic energy and potential energy for an orbiting object for the case n=-2 and for the case n=-3. See Marion and Thornton, Section 6.13 for the form of the virial theorem for this application.

5.6. Modify the inital conditions of the motion in Program 5.4 to demonstrate energy loss for an object passing ahead of a large moving object. Hint: Run Program 5.4 and find the values of velocity and position of the satellite just before it leaves the screen. Reverse the signs of the velocities of the moon and satellite for initial velocities and use the final positions as initial positions. With these initial conditions, the satellite should return to rest before accelerating again.

5.7. Modify Program 5.5 to display the path of the center of mass of the pair of interacting objects. Verify that the center of mass of the system moves in a straight line at constant velocity.

5.8. a. Modify the initial conditions of Program 5.5 to produce bounded orbits with a stationary center of mass located in the center of the screen.
b. Modify the program to display the angular momentum of the system relative to the center of mass for each successive calculation.
c. Add the effect of a small force ("solar wind") of constant magnitude and direction acting on the orbiting objects. A similar effect can be observed in Figure 5.6 due to the effect of an electric field on the path of an electron orbiting in a magnetic field.

5.9. Modify Program 5.5 to plot the path of the particles when a small frictional force proportional to velocity is present in addition to the gravitational force.

5.10. Modify Program 5.6 to calculate the time average of the x-component of velocity and the y-component of velocity. Verify that the time average of the y-component of velocity is equal to E/B when averaged over a complete cycle of the motion. (See Marion and Thornton, Problem 2.16.)

Problems

5.1. a. Write a program to analyze the motion of a projectile in a resisting medium. (See Marion and Thornton, Example 2.7.) Reproduce Figure 2.8 of the Marion and Thornton text using this program.
b. Modify the program to account for the effect of wind directed horizontally. Adjust the wind velocity, drag, and initial conditions of the motion so that the projectile returns to the point from which it was launched.
c. Modify the program so that the friction force is proportional to v^2. The direction of the friction force can be assigned using $-SGN(v)$ as in Problem 3.12.

5.2. a. Program 5.3 is based on equations of motion which reduce the motion of two bodies to the equivalent one-body problem. Orbits are then drawn from the point of view of an observer obove the plane of the orbits and at rest with respect to one of the interacting bodies (mass M). Apply a Galilean transformation to the results of Program 5.3 to plot the path of the moon as seen by an observer located above the plane of the orbit but at rest with respect to the sun. (The path of the moon is wavelike in this case!) To apply the transformation, modify the program to plot the point ($x_1 = x-vt$, $y_1 = y$) instead of the point (x,y). In these transformation equations, v is the apparent velocity of the sun relative to the earth. (See Marion and Thornton, Section 14.2.)

5.3. a. Write a program using the last point approximation to solve the restricted three body problem with gravitational forces. Consider two large objects to be at rest (actually orbiting their mutual center of mass) with a much less massive object moving in response to the gravitational forces.
b. Find initial conditions of the motion which causes "figure-eight" shaped orbits. Hint: see Cromer, *American Journal of Physics*, May, 1981, p. 455.

5.4. a. Write a program to solve the three-body problem for motion in a plane with gravitational forces.
b. Plot the path of the center of mass of the system.
c. Modify the program so that the paths of the particles "bounce" when they encounter the edge of the screen.

5.5. a. Use Lagrange's equation to derive the equations of motion for a simple plane pendulum for which the length of the pendulum changes at a constant rate.
b. Write a program to solve these equations and plot a graph of the angular position of the pendulum as a function of time.
c. Display the value of the total mechanical energy of the pendulum for each successive oscillation.

5.6. a. Use Lagrange's equation to derive the equation of motion for a bead free to slide frictionlessly on a wire ring rotating at constant angular velocity about a vertical axis corresponding to a diameter of the wire ring.
b. Use the last point approximation to solve the equation of motion of the bead and graph the angular position of the bead as a function of time.
c. Alter the equation of motion and solve for the case in which the ring undergoes uniform angular acceleration.

5.7. Using Figure A.2 and the program created in Exercise A.14, write a program to graph the motion of the projectile for which the resisting force is given (See Problem 4.4). Use Euler's method as in Problem 5.1. Instead of using an analytical expression, determine the value of the resisting force using the array created in Exercise A.14 (based on Figure A.2). Interpolate between successive values of force stored in the array in order to improve the accuracy of the results.

5.8. a. Using Lagrange's equation, derive the equations of motion for a planar double pendulum. (A simple plane pendulum of mass m_2 and length l_2 suspended below a simple plane pendulum of mass m_1 and length l_1.)
b. Write a computer program to plot the path of the mass m_2. (To use this problem as an introduction to the topic of deterministic chaos, see the film *The Planar Double Pendulum* produced by Institut für den Wissenschaftlichen Film, Göttingen.)

CHAPTER 6

Motion In Three Dimensions

6.1 Introduction

Perspective drawings of parametric functions in three dimensions were created using programs developed in Chapter 2. This technique can be combined with the numerical methods developed in Chapter 3 to create perspective drawings of the paths of particles moving in three dimensions when the equations of motion and the initial conditions of the motion are known.

Program 2.5 can be altered to illustrate the motion of particles in three dimensions. The angles φ, θ, and ψ are specified as before to provide an arbitrary view of the path of the particle. The graph of the motion of the particle can be scaled to fit the square, flat grid which the program draws in the X-Y plane. By scaling the position of the particle so that the X and Y coordinates of the position remain within the range of the grid, the path of the particle can be conveniently confined to the area viewed on the screen.

Program 6.1 illustrates the general form of programs of this type. The screen display is defined as in Program 2.5. As in the programs of chapters 4 and 5, program lines 500-1000 are reserved to specify the initial conditions of the motion and other parameters required to evaluate the equations of motion. Program lines 1000-2000 are reserved for successive calculations of the velocity and position using Euler's method or one of the associated techniques described in chapter 3. After scaling the coordinates of the particle's position found in each of the successive calculations, the rotation subroutine is used as before to project the points onto the screen.

6.2 Motion of a Projectile in a Rotating Coordinate System

Motion is intuitively related to inertial reference frames. However, motion such as that near the earth's surface is often viewed relative to rotating coordinate systems. To an observer in a rotating coordinate system, the effective force acting on a particle of mass m is given by Eq. 6.1, where ar represents the acceleration relative to the rotating system and af represents the acceleration relative to the fixed system. (See Marion and Thornton, Section 9.3.)

$$F_{eff} = ma_r = ma_f - m\omega \times (\omega \times r) - 2m\omega \times v - m\dot{\omega} \times r \qquad (6.1)$$

In this equation, the origin of the rotating coordinate system is considered to be located on the axis of rotation. If the origin of the rotating coordinate system experiences no acceleration, maf represents the "real" force applied to the moving particle. The other terms in Eq. 6.1 represent "fictitious" forces affecting the motion observed relative to the rotating coordinate system. The centrifugal force is $-m\omega \times (\omega \times r)$. The Coriolis force acting on the moving particle is $-2m\omega \times v$. The last term in the equation (which seems to have no name) is often negligible as it represents the influence of the angular acceleration $\dot{\omega}$ of the rotating coordinate system.

In the case in which the z-axis is the axis of rotation of the rotating system, ω is a vector in the z direction.

$$\omega = \omega e_z \qquad (6.2)$$

Using cartesian coordinates, the components of the applied force become F_x, F_y, and F_z. Equation 6.1 can then be written as a set of coupled equations.

$$a_x = \omega^2 x + 2\omega \dot{y} + \dot{\omega} y + F_x/m \qquad (6.3)$$

$$a_y = \omega^2 x - 2\omega \dot{x} - \dot{\omega} x + F_y/m \qquad (6.4)$$

$$a_z = F_z/m \qquad (6.5)$$

It is important to note that all of the coordinates and velocities in this set of equations are defined relative to the rotating coordinate system.

If the only applied force is a uniform gravitational force in the -z direction, the set of equations is simplified further by substituting:

$$F_x = 0 \ ; \ F_y = 0 \ ; \ F_z = -mg \qquad (6.6)$$

The resulting set of coupled equations are the equations of motion for an object moving in a coordinate system rotating about the z axis with a uniform gravitational force acting in the -z direction.

$$a_x = \omega^2 x + 2\omega \dot{y} + \dot{\omega} y \qquad (6.7)$$

$$a_y = \omega^2 x - 2\omega \dot{x} - \dot{\omega} x \qquad (6.8)$$

$$a_z = -g \qquad (6.9)$$

Program 6.1 numerically solves the set of equations above and creates a perspective drawing of the path of the particle. The angular acceleration ($\dot{\omega}$) and the initial angular velocity (ω_0) of the rotating system are specified in line number 505 and line number 510 respectively. Line number 530 specifies the initial position of the particle in the rotating system. The initial components of velocity relative to the rotating system are specified in line number 540.

Successive calculations are performed for each time increment dt. Line number 1015 calculates the value of ω, the angular velocity of the rotating system at time t using the relation; $\omega = \omega_0 + \dot{\omega} t$. This value is then used with Eq. 6.7, Eq. 6.8, and Eq. 6.9 to determine the new values of the velocity components (line numbers 1020-1040) and position (line numbers 1050-1070). The position of the particle is scaled to the screen in line number 1080. In this particular example, each square in the grid forming the x-y plane represents a square with sides 100 meters long.

The position of the particle is then projected onto the screen using the rotation subroutine. The screen display represents the motion as seen by an observer at rest with respect to the rotating coordinate system.

```
9   REM     PROGRAM (6.1)
10  REM     THIS PROGRAM USES EULER'S ANGLES FOR ROTATION
80  REM
                ***** SET UP GRAPHICS CHARACTERISTICS *****


90  &  HGR2 : &  HCOLOR= 15: & B COLOR= 0: &  PRINT : & MODE(1)
95  REM
                ***** SET UP SCREEN DISPLAY *****

100 REM       ANGLE A1 IS PHI:REM ANGLE A2 IS THETA:REM ANGLE A3 IS
PSI:REM ALL ANGLES ARE EXPRESSED IN RADIANS
110 A1 = 1.57
120 A2 = 1.57
130 A3 = 0
135 &  GOTO 510,10: PRINT "A1=";A1
137 &  GOTO 510,18: PRINT "A2=";A2
139 &  GOTO 510,26: PRINT "A3=";A3
140 S1 =  SIN (A1):C1 =  COS (A1)
150 S2 =  SIN (A2):C2 =  COS (A2)
```

6.2 MOTION OF A PROJECTILE IN A ROTATING COORDINATE SYSTEM

```
160 S3 =  SIN (A3):C3 =  COS (A3)
170  REM                                              L1  L2  L3
180  REM   CALCULATE ELEMENTS OF TRANSFORMATION MATRIX L4  L5  L6
190  REM                                              L7  L8  L9

200 L1 = C3 * C1 - C2 * S1 * S3
210 L2 =  - S3 * C1 - C2 * S1 * C3
220 L3 = S2 * S1
230 L4 = C3 * S1 + C2 * C1 * S3
240 L5 =  - S3 * S1 + C2 * C1 * C3
250 L6 =  - S2 * C1
260 L7 = S3 * S2
270 L8 = C3 * S2
280 L9 = C2
290 D = 20: REM           DISTANCE OF VIEWER FROM SCREEN (CM)
300 XV = D * L1
310 YV = D * L2
320 ZV = D * L3
330  READ A,X1,Y1,Z1,X,Y,Z
340  IF A = 16 GOTO 500
350  GOSUB 2000
390 SX = XS:SZ = ZS:X = X1:Y = Y1:Z = Z1
395  GOSUB 2000
400  &  HPLOT XS,ZS TO SX,SZ
410  IF A = 4 THEN  &  GOTO XS,ZS: PRINT "X"
420  IF A = 11 THEN  &  GOTO XS,ZS: PRINT "Y"
430  IF A = 15 THEN  &  GOTO XS,ZS: PRINT "Z"
440  GOTO 330
500  REM
              ***** SET UP INITIAL CONDITIONS *****

505 VW = 0
510 W0 = .1
520 G = 9.8
530 X5 = 300:Y5 = 300:Z5 = 0
540 V1 = -50:V2 = -50:V3 = 50
550 DT = .1
1000  REM
        *****   CALCULATE AND PLOT VELOCITY AND POSITION  *****

1010  FOR T = 0 TO 23 STEP DT
1015 W = W0 + VW * T : REM ANGULAR VELOCITY
1020 V4 = V1 + (W * W * X5 + 2 * V2 * W + Y5 * VW) * DT
1030 V5 = V2 + (W * W * Y5 - 2 * V4 * W - X5 * VW) * DT
1040 V6 = V3 - G * DT
```

```
1050 X6 = X5 + V4 * DT : REM  LAST POINT APPROXIMATION
1060 Y6 = Y5 + V5 * DT
1070 Z6 = Z5 + V6 * DT
1080 X = X5 / 100:Y = Y5 / 100:Z = Z5 / 100
1090  GOSUB 2000
1095 &  HPLOT XS,ZS
1100 V1 = V4:V2 = V5:V3 = V6
1110 X5 = X6:Y5 = Y6:Z5 = Z6
1120  NEXT T
1900  REM               ***** ROTATION SUBROUTINE *****

2000 XO = X - XV: REM           TRANSLATION OF COORDINATES TO MOVE
OBSERVER TO ORIGIN OF COORDINATES
2010 YO = Y - YV
2020 ZO = Z - ZV
2025  REM      ***** APPLY ROTATION MATRIX  ****
2030 X3 = L1 * XO + L2 * YO + L3 * ZO
2040 Y3 = L4 * XO + L5 * YO + L6 * ZO
2050 Z3 = L7 * XO + L8 * YO + L9 * ZO
2060  REM      ***** PROJECT ROTATED OBJECT ON SCREEN *****
2070 XS = 280 + 40 * D * Y3 / ( - X3)
2080 ZS = 96 - 20 * D * Z3 / ( - X3)
2090  RETURN
3000  REM               ***** DATA FOR SCREEN DISPLAY *****

3010  DATA   1,3,3,0,-3,3,0
3020  DATA   2,3,2,0,-3,2,0
3030  DATA   3,3,1,0,-3,1,0
3040  DATA   4,4,0,0,-4,0,0
3050  DATA   5,3,-1,0,-3,-1,0
3060  DATA   6,3,-2,0,-3,-2,0
3070  DATA   7,3,-3,0,-3,-3,0
3080  DATA   8,3,3,0,3,-3,0
3090  DATA   9,2,3,0,2,-3,0
3100  DATA   10,1,3,0,1,-3,0
3110  DATA   11,0,4,0,0,-4,0
3120  DATA   12,-1,3,0,-1,-3,0
3130  DATA   13,-2,3,0,-2,-3,0
3140  DATA   14,-3,3,0,-3,-3,0
3150  DATA   15,0,0,4,0,0,0
3160  DATA   16,0,0,0,0,0,0
5000  END
```

6.2 MOTION OF A PROJECTILE IN A ROTATING COORDINATE SYSTEM

Before considering the motion of a projectile, it is useful to consider the force-free motion of a stationary particle as viewed in a rotating coordinate system. This can be done by changing the value of g (line number 520) and the value of initial velocity of the particle (line number 540) to zero. In this case, no applied forces affect the motion of the particle.

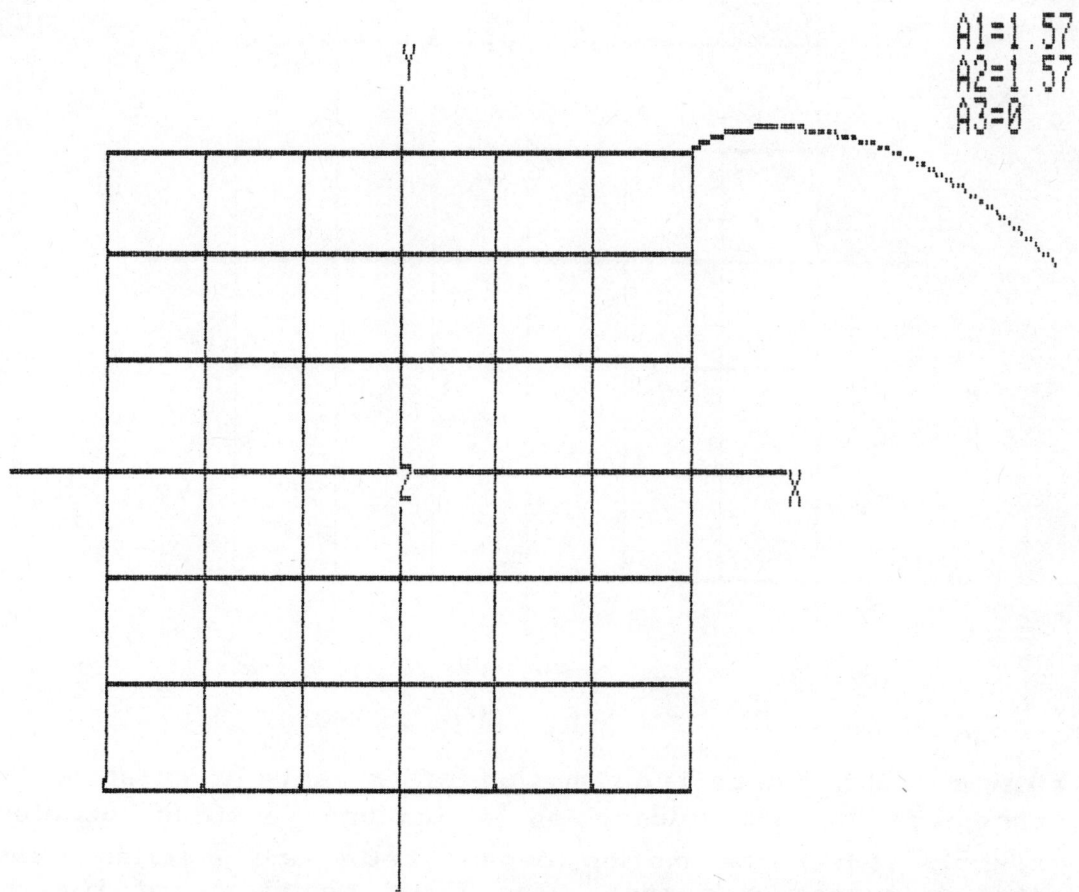

Figure 6.1. Force-free motion of a particle relative to a rotating coordinate system (ω=1 rad/s). In this figure, g=0. In addition, the system rotates at a constant angular velocity ($\dot{\omega}$=0). As in Chapter 2, the values of A1, A2, and A3 represent the respective values of the Eulerian angles, φ, θ, and ψ. The particle was initially at rest with respect to the rotating coordinate system and was located on the upper right corner of the grid (the point 3,3).

In Figure 6.1 the influence of the centrifugal force and the Coriolis force can be clearly observed. The centrifugal force causes the particle to accelerate outward from the axis of rotation (z-axis) while the Coriolis force causes a deflection perpendicular to the motion of the particle.

However, in Figure 6.2, the influence of the centrifugal force seems to be cancelled by the influence of the angular acceleration $\dot{\omega}$ of the rotating coordinate system causing the particle to remain at a constant distance from the origin. The particle merely moves with increasing velocity in a circular path around the z-axis.

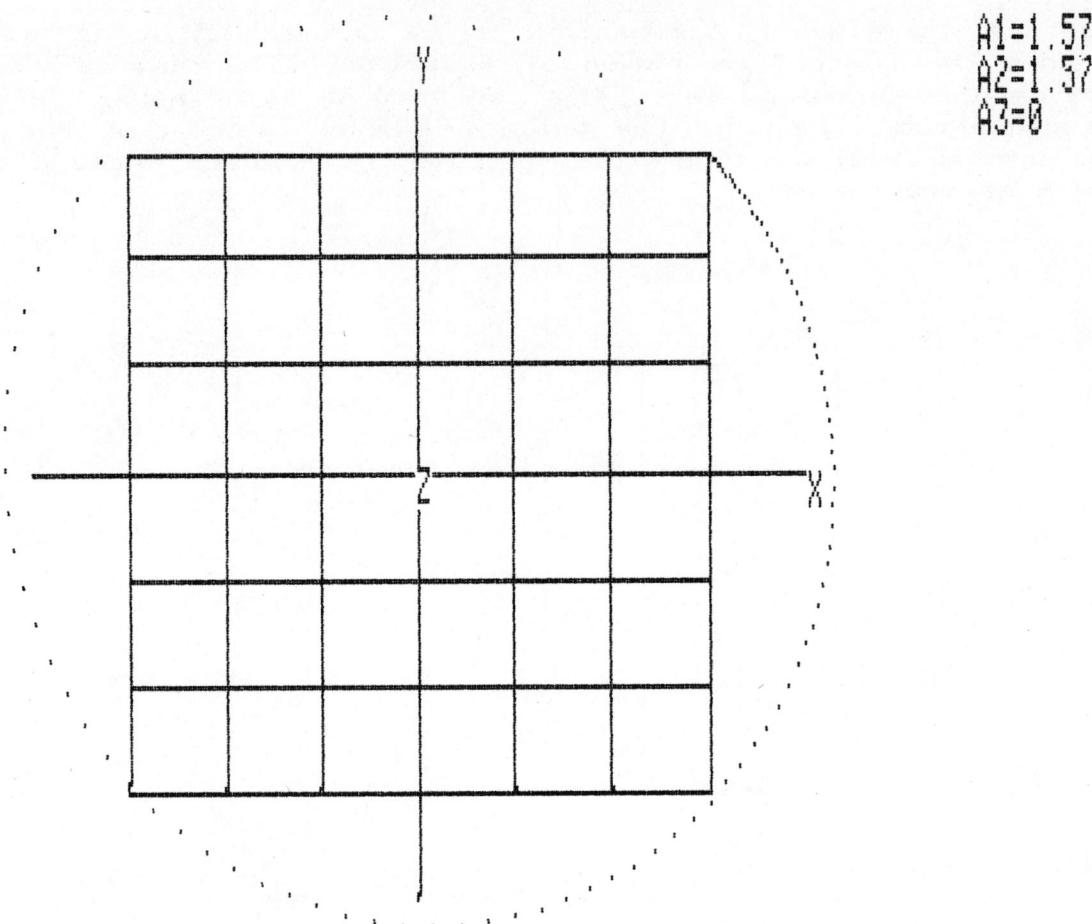

Figure 6.2. Force-free motion of a particle relative to a rotating coordinate system undergoing a uniform angular acceleration ($\ddot{\omega}=.1$ rad/s^2). When the motion began (t=0), the angular velocity of the rotating system was zero. The initial conditions of the motion of the particle were the same as those for Figure 6.1.

   The difference in the nature of the motion in the two examples can be understood by carefully considering the initial conditions of the motion in each case. In Figure 6.2, both the inertial (non-rotating) system and the rotating system were initially fixed and the particle was at rest in both systems. As the angular velocity of the rotating system (the screen) began to increase, the particle began to move in a circular path with an angular velocity opposite that of the rotating coordinate system. However, in the inertial system, the particle remained at rest with no applied forces. The path of a particle which is motionless in the inertial system is thus a circle when viewed from the rotating system.
   The path of the particle in Figure 6.1 can be understood in a similar fashion. The particle was initially at rest in the rotating coordinate system but was initially moving in the inertial system. As the particle continued to move in a straight line at constant velocity relative to the inertial system, the path observed relative to the computer

6.2 MOTION OF A PROJECTILE IN A ROTATING COORDINATE SYSTEM

screen (the rotating coordinate system) was the path shown in Figure 6.1. The concept of the "fictitious" centrifugal force and Coriolis force is used to make the motion observed in the rotating coordinate system be consistent with Newton's Second Law for an inertial coordinate system.

Force-free motion was displayed in Figure 6.1 and in Figure 6.2 when the value of the acceleration due to gravity (g) was zero. Projectile motion occurs when g is given a positive value causing the moving particle to undergo uniform acceleration in the -z direction. Figure 6.3 illustrates the motion of a projectile in a coordinate system rotating at constant angular velocity ω. The path of the projectile appears to undergo a deflection across the x-y plane due to the action of the Coriolis force. Figure 6.4 illustrates the motion of a similar projectile relative to a coordinate system undergoing uniform angular acceleration. The (nameless) last term in Eq. 6.1 is added to the Coriolis force to create the complex motion observed.

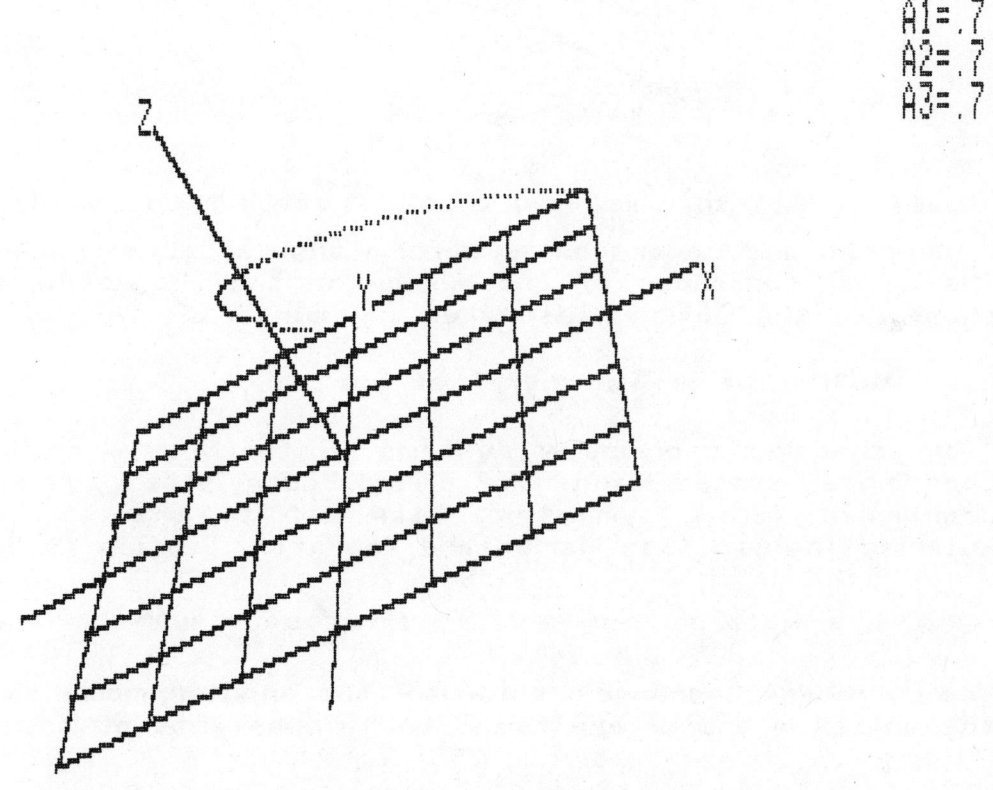

Figure 6.3. The motion of a projectile relative to a rotating coordinate system is shown. The initial conditions of the motion are given in Program 6.1.

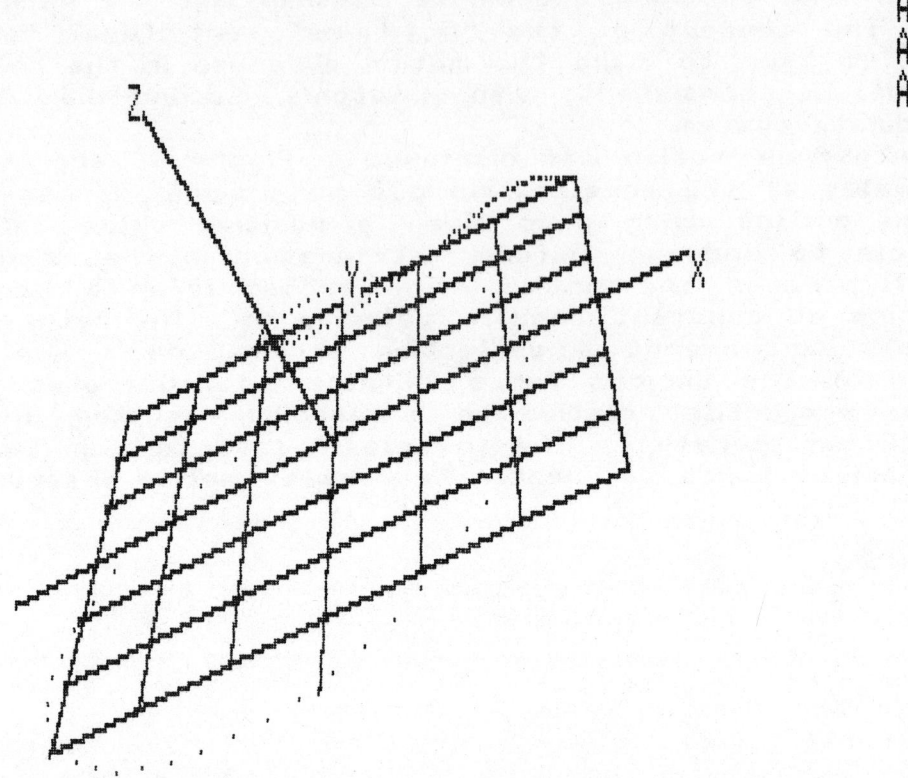

Figure 6.4. In this figure, the rotating coordinate system starts from rest and undergoes a uniform angular acceleration ($\ddot{\omega}=.1$ rad/s^2). The initial conditions of the motion of the projectile are the same as those for the motion illustrated in Figure 6.3.

6.3 Motion of a Symmetrical Top

By choosing coordinate systems in which the origins of the body coordinate system and the fixed coordinate system coincide, the Lagrangian for a symmetrical top can be written in terms of the Eulerian angles. (See Marion and Thornton, Section 10.10.)

$$L = I_{12}(\dot{\varphi}^2\sin^2\theta + \dot{\theta}^2)/2 + I_3(\dot{\varphi}\cos\theta + \dot{\psi})^2/2 - Mgh\cos\theta \qquad (6.7)$$

By applying Lagrange's equation, the angular momenta associated with the angles φ and ψ are found to be constants of the motion.

$$P_\varphi = (I_{12}\sin^2\theta + I_3\cos^2\theta)\dot{\varphi} + I_3\dot{\psi}\cos\theta = \text{constant} \qquad (6.8)$$

$$P_\psi = I_3(\dot{\psi} + \dot{\varphi}\cos\theta) = \text{constant} \qquad (6.9)$$

These equations can be solved for $\dot{\varphi}$ and $\dot{\psi}$ in terms of θ and the constants of the motion, P_φ and P_θ.

6.3 MOTION OF A SYMMETRICAL TOP

$$\dot{\varphi} = P_\varphi - P_\psi \cos\theta / I_{12} \sin^2\theta \qquad (6.10)$$

$$\dot{\psi} = P_\psi / I_3 - (P_\varphi - P_\psi \cos\theta)\cos\theta / I_{12} \sin^2\theta \qquad (6.11)$$

Lagrange's equation for the coordinate θ yields an equation for $\ddot{\theta}$ in terms of the angle θ and the angular velocities $\dot{\varphi}$ and $\dot{\theta}$.

$$\ddot{\theta} = (((I_{12} - I_3)\dot{\varphi}^2 \cos\theta - I_3 \dot{\psi}\dot{\varphi} + mgh)\sin\theta)/I_{12} \qquad (6.12)$$

Program 6.2 uses the last-point approximation to calculate the angular position of the axis of symmetry of the top as a function of time. The values of the angular momenta P_φ and P_ψ are calculated in line number 590 and line number 600. These constants of the motion are used in the subsequent calculations of $\dot{\psi}$ (line number 1020), $\dot{\varphi}$ (line number 1030), and $\dot{\theta}$ (line number 1040). The values of the Eulerian angles φ, θ, and ψ which determine the spatial orientation of the top are then calculated in line number 1060, line number 1070, and line number 1080, respectively.

The cartesian coordinates of the tip of the axis of symmetry are determined in line number 1080, line number 1090, and line number 1100. In order to fit the screen grid conveniently, the axis of symmetry is considered to have a length of 3 cm. The location of the tip of the axis of rotation is then projected onto the screen using the rotation subroutine.

The resulting figure traces the path of the axis of symmetry of the top on a spherical surface (with a radius of 3 cm). The general form of the figure depends on the initial conditions of the motion and the structural details of the top. (See Marion and Thornton, Section 10.10.)

```
5   REM PROGRAM (6.2)
10  REM   THIS PROGRAM USES EULER'S ANGLES FOR ROTATION
80  REM
              ***** SET UP GRAPHICS CHARACTERISTICS *****


90  &  HGR2 : &  HCOLOR= 15: & B COLOR= 0: &  PRINT : & MODE(1)
95  REM
              ***** SET UP SCREEN DISPLAY *****


100 REM       ANGLE A1 IS PHI:REM ANGLE A2 IS THETA:REM ANGLE A3 IS
PSI:REM ALL ANGLES ARE EXPRESSED IN RADIANS
110 A1 = 0
120 A2 = 0
130 A3 = 0
135 &  GOTO 510,10: PRINT "A1=";A1
137 &  GOTO 510,18: PRINT "A2=";A2
139 &  GOTO 510,26: PRINT "A3=";A3
```

78 · 6 · MOTION IN THREE DIMENSIONS

```
140 S1 =  SIN (A1):C1 =  COS (A1)
150 S2 =  SIN (A2):C2 =  COS (A2)
160 S3 =  SIN (A3):C3 =  COS (A3)
170  REM                                                L1  L2  L3
180  REM   CALCULATE ELEMENTS OF TRANSFORMATION MATRIX  L4  L5  L6
190  REM                                                L7  L8  L9
200 L1 = C3 * C1 - C2 * S1 * S3
210 L2 =  - S3 * C1 - C2 * S1 * C3
220 L3 = S2 * S1
230 L4 = C3 * S1 + C2 * C1 * S3
240 L5 =  - S3 * S1 + C2 * C1 * C3
250 L6 =  - S2 * C1
260 L7 = S3 * S2
270 L8 = C3 * S2
280 L9 = C2
290 D = 20: REM           DISTANCE OF VIEWER FROM SCREEN (CM)
300 XV = D * L1
310 YV = D * L2
320 ZV = D * L3
330  READ A,X1,Y1,Z1,X,Y,Z
340  IF A = 16 GOTO 500
350  GOSUB 2000
390 SX = XS:SZ = ZS:X = X1:Y = Y1:Z = Z1
395  GOSUB 2000
400 &  HPLOT XS,ZS TO SX,SZ
410  IF A = 4 THEN  &  GOTO XS,ZS: PRINT "X"
420  IF A = 11 THEN  &  GOTO XS,ZS: PRINT "Y"
430  IF A = 15 THEN  &  GOTO XS,ZS: PRINT "Z"
440  GOTO 330
500  REM
                        ***** SET UP INITIAL CONDITIONS *****

505 X0 = SX:Z0 = SZ
510 SI = 0:VS = 100
520 PH = 0:VP = 6: REM       PH IS PHI AND VP IS ANGULAR VELOCITY
ASSOCATED WITH PHI
530 TH = .6:VT = 6
540 I3 = 1000: REM   MOMENT OF INERTIA
550 I2 = 3000:I1 = 3000
560 M = 100
570 H = 5
580 G = 980
590 PP = ((I2 * ( SIN (TH) ^ 2) + I3 * ( COS (TH) ^ 2)) * VP) + I3
 *  COS (TH) * VS
```

6.3 MOTION OF A SYMMETRICAL TOP

```
600 PS = I3 * (VS + VP *  COS (TH))
610 DT = .005
1000  REM
        *****    CALCULATE AND PLOT VELOCITY AND POSITION   *****

1010  FOR T = 0 TO 10 STEP DT
1020 VS = (PS / I3) - ((PP - PS *  COS (TH)) *  COS (TH)) / (I1 * (
 SIN (TH) ^ 2))
1030 VP = ((PP) - (PS *  COS (TH))) / (I1 * ( SIN (TH) ^ 2))
1040 V1 = VT + ((((I1 - I3) * (VP ^ 2) *  COS (TH) - I3 * VS * VP +
 M * G * H) *  SIN (TH)) / I1) * DT
1060 P1 = PH + VP * DT
1070 T1 = TH + V1 * DT: REM    LAST POINT APPROXIMATION
1072 SI = S1:PH = P1:TH = T1:VT = V1
1080 X = 3 * ( SIN (TH) *  COS (PH))
1090 Y = 3 * ( SIN (TH) *  SIN (PH))
1100 Z = 3 *  COS (TH)
1110  GOSUB 2000
1120  &  HPLOT XS,ZS
1140  NEXT T
1900  REM
                 ***** ROTATION SUBROUTINE *****


2000 XO = X - XV: REM              TRANSLATION OF COORDINATES TO MOVE
 OBSERVER TO ORIGIN OF COORDINATES
2010 YO = Y - YV
2020 ZO = Z - ZV
2025  REM     ***** APPLY ROTATION MATRIX ****
2030 X3 = L1 * XO + L2 * YO + L3 * ZO
2040 Y3 = L4 * XO + L5 * YO + L6 * ZO
2050 Z3 = L7 * XO + L8 * YO + L9 * ZO
2060  REM    ***** PROJECT ROTATED OBJECT ON SCREEN *****
2070 XS = 280 + 40 * D * Y3 / ( - X3)
2080 ZS = 96 - 20 * D * Z3 / ( - X3)
2090  RETURN
3000  REM
              ***** DATA FOR SCREEN DISPLAY *****

3010  DATA   1,3,3,0,-3,3,0
3020  DATA   2,3,2,0,-3,2,0
3030  DATA   3,3,1,0,-3,1,0
3040  DATA   4,4,0,0,-4,0,0
3050  DATA   5,3,-1,0,-3,-1,0
3060  DATA   6,3,-2,0,-3,-2,0
```

```
3070 DATA 7,3,-3,0,-3,-3,0
3080 DATA 8,3,3,0,3,-3,0
3090 DATA 9,2,3,0,2,-3,0
3100 DATA 10,1,3,0,1,-3,0
3110 DATA 11,0,4,0,0,-4,0
3120 DATA 12,-1,3,0,-1,-3,0
3130 DATA 13,-2,3,0,-2,-3,0
3140 DATA 14,-3,3,0,-3,-3,0
3150 DATA 15,0,0,4,0,0,0
3160 DATA 16,0,0,0,0,0,0
5000 END
```

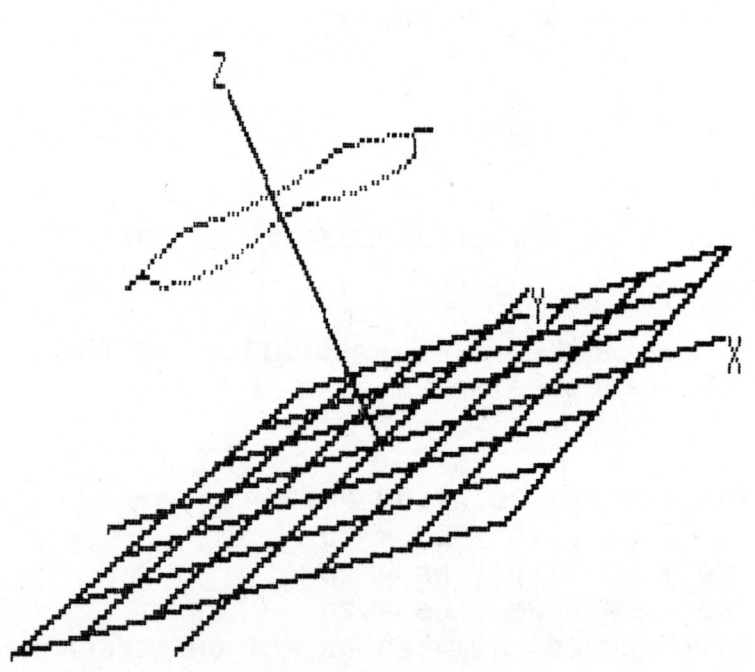

Figure 6.5. Motion of a symmetric top resulting when the initial conditions of the motion are such that φ̇, the angular velocity of precession, is always positive. The amplitude of the oscillation of the angle θ (nutation) can also be varied by changing the initial conditions of the motion.

6.3 MOTION OF A SYMMETRICAL TOP

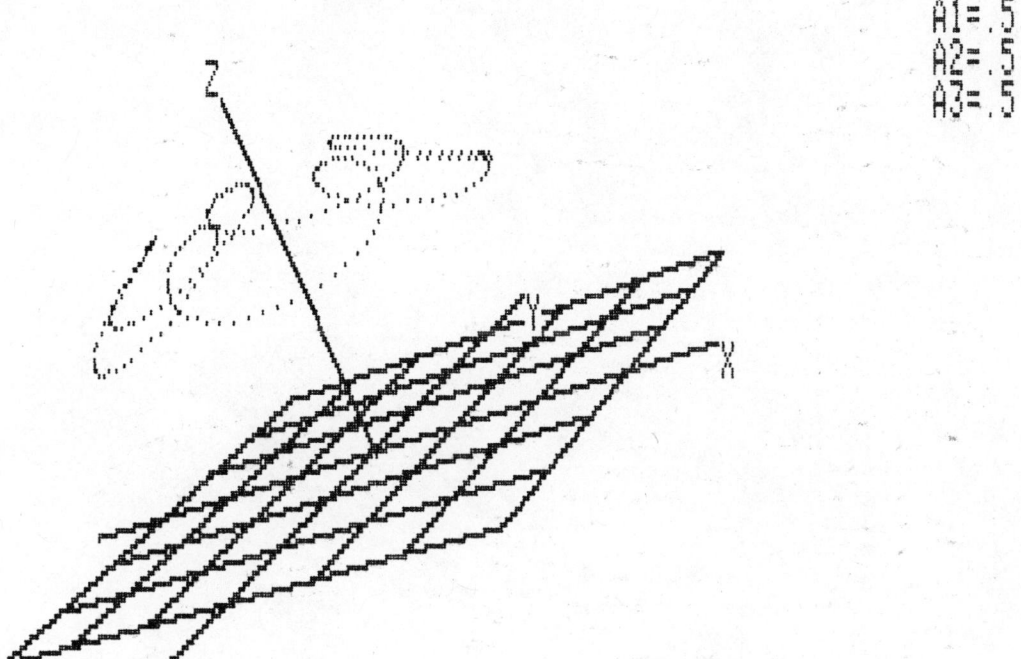

Figure 6.6. In this figure, the looping motion results when the initial conditions are such that the value of $\dot{\varphi}$ is negative when θ is small. The net precessional motion remains in the $+\varphi$ direction.

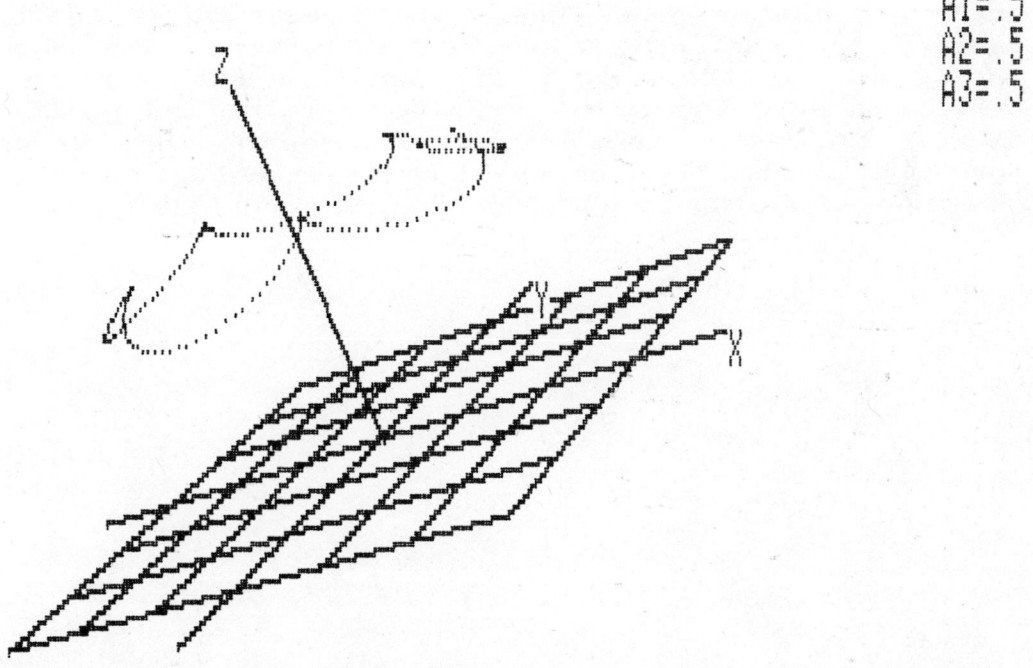

Figure 6.7. This figure represents the usual motion of a top. The top is released from rest such that $\dot{\theta}=0$ and $\dot{\varphi}=0$ but leaning moderately. After release, the top momentarily falls in the gravitational field. The cusplike motion results as the top returns to the initial angle (θ) from which it was released and then repeats the fall as it precesses around the z-axis.

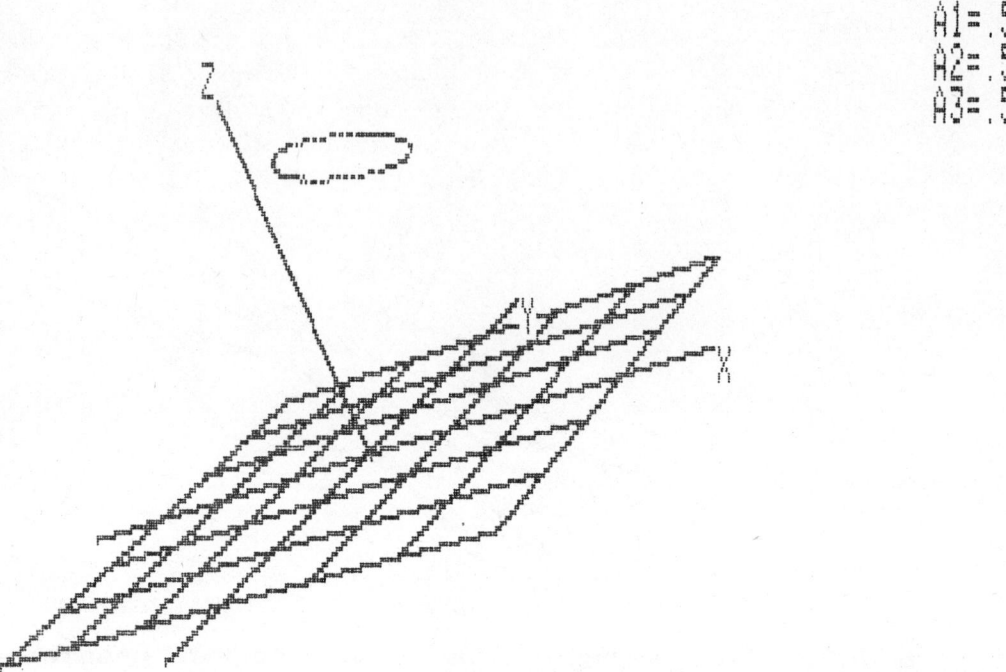

Figure 6.8. When the value of g is zero, the top is said to undergo force-free motion. In this case, the precession rate ($\dot{\varphi}$) is uniform and the nutation disappears. The precession is centered about the direction of the angular momentum vector. The angular momentum vector for the top is thus directed from the origin of the coordinate system through the center of the circle formed by the motion of the axis of the top in this figure. Because no torques act on the top, the angular momentum of the top is a constant both in magnitude and direction. (See Marion and Thornton, Section 10.10.)

Exercises

6.1. Modify the initial conditions of Program 6.1 to analyze the motion of a projectile relative to a rotating coordinate system for which the angular velocity is decreasing. Adjust the angular acceleration of the coordinate system so that the angular velocity of the system reverses while the projectile is in flight.

6.2. Modify Program 6.1 to calculate the deflection of a falling object in a rotating coordinate system. Compare the results with the predictions based on analytical methods. (See Marion and Thornton, Example 9.3, Section 9.4.)

6.3. Modify Program 6.2 to analyze the motion of an inverted symmetrical top. The simplest way to accomplish this is to change the sign associated with the acceleration due to gravity. This type of motion can also be observed in a spinning top constructed with a rounded base. The top will exhibit a rocking motion but will always right itself if the top is not spinning. If friction is present, the "wobble" observed in the motion increases and can invert the spinning top (called a "tippy top").

Problems

6.1. a. Use Lagrange's equation to derive the equations of motion for a spherical pendulum using spherical coordinates. (See Marion and Thornton, Problem 6.31.)
b. Write a program to solve these equations using the last point approximation and display the motion in three dimensions.
c. Find initial conditions for which the path of the pendulum bob is circular and the value of θ is constant. In this case, the pendulum is called a conical pendulum.

6.2. a. Solve the three-body problem with gravitational forces. Display the motion in three dimensions instead of limiting the motion to a plane as in the previous chapter.
b. Modify the program to display the path of the center of mass of the system.
c. Modify the program so that the particles "bounce" when they reach the limits of the region defined by the three dimensional display.

6.3. Create "three-dimensional Lissajous curves" by solving the equation of motion for a three-dimensional harmonic oscillator. Use cartesian coordinates and the last point approximation for this program. See Program 5.1 for the two-dimensional version of this program.

6.4. a. Write a program to analyze the motion of a charged particle in a magnetic field and display the motion in three dimensions.
b. Modify the program to display the effect of adding an electric field to the system. See Program 5.6 for a two-dimensional version of this program. The two-dimensional version is limited to the display of motion in a plane.

6.5. a. Using Lagrange's equation, derive the equation of motion for a "dumbell" shaped rotor for which $m_1 = m_2$ and $|r_1| = |r_2|$. (See Marion and Thornton, Section 10.3, Figure 10.2.) Use two generalized coordinates, one to specify the angle between the shaft and the axis of rotation (θ) and another to specify the angular position of the masses in their plane of rotation (φ).
b. Write a program to plot graphs of θ and φ as a function of time.
c. Display the paths of the masses of the "dumbell" using the three-dimensional plotting technique.

6.6. a. Write a program to analyze the motion of two coupled oscillators by plotting graphs of x_1 and x_2 as a function of time. (For the appropriate equation of motion, see Marion and Thornton, Section 11.2.)
b. Specify initial conditions to verify the characteristic frequencies and normal modes of this system.
c. Extend the equations of motion and the program developed above to apply to the motion of a similar system of three coupled oscillators.

APPENDIX A

Computer Graphics

A.1 Introduction

Programming languages for almost all microcomputers distinguish between text displays and graphics displays. Although text operations in the BASIC programming language have become increasingly standarized in both function and syntax, very little standardization of graphics operations has taken place. Due to the absence of standards, it is not possible to create graphics programs which can be operated without alteration on a wide variety of types of computers. The practical solution to this problem is to develop programs which incorporate only the limited set of graphics operations available to users of almost any type of microcomputer.

The programs of this text thus incorporate only four distinct graphics operations; point-plotting, line-drawing, cursor positioning, and text printing. Users of a particular type of microcomputer can then determine the correct syntax to perform these operations with whatever computer they choose. This appendix outlines and compares these operations in three well known graphics packages; Beagle Graphics for the Apple IIe, DEC ReGIS as used with many products of Digital Equipment Corporation, and GW-BASIC for computers that are compatible with the IBM-pc. Appendix B describes the syntax required to perform graphics operations with each of these systems.

Example programs are written for use with Beagle Graphics, the least expensive of these systems. With the information contained in Appendix B, the programs of the text can easily be converted for use with any of the systems described. If none of these systems are to be used, students can interpret the function of each graphics operation and determine the syntax required to perform the operation using other computers not mentioned here.

A.2 Screen Concepts

When used to display graphics, the screen of the computer monitor can be thought of as a cartesian coordinate system. The screen contains a finite number of picture elements (pixels) each of which is designated by a position (x,y) in this coordinate system. Images are formed by controlling the display of arrays of pixels. Precision of the images is determined by the number of visibly distinct pixels which can be displayed on the screen. The larger the number of pixels, the greater the detail of images created on the screen.

For most computers, the position of the pixel located at the upper left corner of the screen is the origin of the coordinate system; the point (0,0). The positions of pixels located to the right of the origin are represented by increasing values of the x coordinate. For example, using Beagle Graphics, the position of the pixel located at the upper right corner of the screen is designated (559,0).

The positions of pixels located below the origin are represented by increasing values of the y coordinate. In the Beagle Graphics system the pixel at the lower left corner of the screen is labelled (0,191) and the position of the pixel at the lower right corner is thus (559,191). All pixels on the screen are labelled in this way. For example, the position of the pixel nearest the center of the screen is (280,96). The screen contains an array of 560 by 192 positions at which a pixel can be displayed.

In addition to precision and location, pixels are characterized by aspect ratio: the ratio of height to width. The aspect ratio of pixels must be considered when creating graphics images in order to minimize distortion. When using a computer with a correctly adjusted monitor, the aspect ratio of pixels of the Beagle graphics system is two. A horizontal line thus requires twice as many pixels as a vertical line of equal length when this system is used. A "square" with dimensions of 20 pixels x 20 pixels will appear to be a rectangle. To appear on the screen as a square, the figure must be 40 pixels wide and 20 pixels high.

The aspect ratio for other graphics systems may be different than the value used in the example above. For example, the aspect ratio for IBM-pc compatible computers is 2.25 when GW-BASIC is used. In addition, the aspect ratio may be affected by the monitor used to display graphics. If the monitor cannot be adjusted to display undistorted images when the graphic figures are created with the correct theoretical aspect ratio, a working value of the aspect ratio may be determined by trial and error.

A.3 Graphics Operations

The programs of this text incorporate only the graphics operations described below. The syntax shown is specific to the Beagle Graphics system used in the development of these programs.

A.3 GRAPHICS OPERATIONS

1. Point-plotting (Display a single pixel)

In many operations such as plotting graphs, it is useful to cause the pixel at a single screen position to be displayed. To display the pixel at the position (50,50), the statement &HPLOT 50,50 is executed. If the position is to be specified by variables x and y defined earlier in the program, the variables are incorporated in the statement. The statement then becomes &HPLOT X,Y.

A straight line can be drawn (slowly!) by repeated execution of the point-plotting statement. For example, when incorporated into a graphics program the following program line uses point-plotting to draw a straight line corresponding to the equation y = 2x for values of x from 0 and 50 by displaying individual pixels.

    100 FOR X=0 to 50:Y=2*X:&HPLOT X,Y :NEXT X

2. Line-drawing (Draw a straight line)

Instead of using point-plotting, a faster and simpler way to draw the straight line in is to make use of the line-drawing statement &HPLOT 0,0 to 50,100. Execution of this statement almost instantly draws a straight line from the point (0,0) to the point (50,100). As in the previous example, variables can be incorporated in the statement.

3. Positioning the graphics cursor

Before executing certain graphics operations (such as displaying text), the graphics cursor must first be directed to the desired position on the screen. For example, the statement which directs the graphics cursor to the center of the screen (position 280,96) is &GOTO 280,96. If variables x and y have been specified earlier in the program, the execution of the statement &GOTO X,Y directs the cursor to the position (x,y).

4. Printing text on graphics screen

Before printing text on the screen, the graphics cursor must first be directed to the screen location at which text is to be printed. The BASIC language PRINT statement will then operate in the usual manner but with the first printed character positioned at the location specified. See line numbers 340-390 and line numbers 1120 to 1130 of Program 4.1 for examples of the use of this statement. The statements in these program lines are used to first label the coordinate axes of the figure and then to continuously display the values of velocity and position of the moving object as the calculations proceed.

A.4 Graphing a Mathematical Function
(Cartesian Coordinates)

Equation A.1, which results from the analytical solution of the equation of motion of an object moving horizontally through a retarding medium (Eq. 4.4) is used here to provide an example of plotting techniques. The equation describes the position of the moving object as a function of time. The values of v_0 and b are constants determined by the initial conditions of the motion and by the nature of the frictional forces.

$$x = v_0(1 - e^{-kt})/k \qquad (A.1)$$

The program developed below plots the position (vertical axis) as a function of time (horizontal axis) over the time interval from 0 to 10 seconds. The graph is nearly identical to the upper curve of Figure 4.1.

In order to plot the function, it is first necessary to specify the values of the constants; v0 and k. In a BASIC language program, the steps below are used to specify the values of constants. Each statement is numbered and is to be executed in the sequence defined by these numbers. The instruction REM in each statement (preceded by a colon to indicate that is a sparate procedure) contains remarks for the benefit of the programmer and does not affect the execution of the program.

```
100 V0=10 :REM INITIAL VELOCITY
110 k=5   : REM DRAG COEFFICIENT
```

After specifying the values of constants, a procedure called a for-next loop is then used to calculate successive values of the function over the time interval of interest. The for-next loop first directs the computer to calculate the value of the function at the initial time specified. (t=0 in this case.) In the loop shown below, the value of X is printed and the calculation is then repeated after time is advanced incrementally by the amount specified by the STEP instruction. (STEP .1 in the statements below.) This process is repeated until the function has been evaluated for each of the specified values of time over the entire range defined in the loop. (0 to 10.)

```
120 FOR T=0 TO 10 STEP .1
 130 X=V0*(1-EXP(-K*T))/K :REM CALCULATIONS OCCUR HERE SEE EQ. A.1
190 PRINT X
200 NEXT T
```

Now that values of X can be calculated, the computer may be instructed to plot the coordinates corresponding to X and T instead of printing the values. If the print statement of line number 190 is replaced by the point-plotting statement which displays the pixel (T,X), the program will create a small, inverted graph of the function.

A.5 Scaling and Positioning the Figure

In order to produce a satisfactory graph of the function, the figure must first undergo a change of scale to enlarge its dimensions and then be inverted to provide the correct orientation. Because the original figure is too small, the values of t and x are multiplied by scale factors greater than one in order to enlarge the graph. The scaled coordinates (called (TS,XS) below) are then plotted. To perform these operations, line number 190 of the program above is deleted and replaced by the program lines shown below.

```
170 TS=50*T  :REM HORIZONTAL SCREEN LOCATION
180 XS=9*X   :REM VERTICAL SCREEN LOCATION
190 &HPLOT(TS,XS)
```

By using a scale factor of 50 in line number 170, the horizontal dimension of the graph is enlarged to 50 times its original size. The scale factor of nine used in line number 180 enlarges the vertical dimension to nine times its original size. The graph is enlarged by this change but the figure is still inverted.

The first step in righting the figure is to translate the figure down the screen a distance of 90 pixels in order to allow room for the upright figure. In order to accomplish this translation, the statement in line number 180 becomes: XS = 90 + 9*X. The simple expedient of changing the sign of the scale factor is then used to compensate for the fact that the computer screen reverses the usual convention of cartesian coordinates by designating positions lower on the screen with larger values of the screen coordinate. To plot the function upright the statement in line number 180 is finally: XS = 90 - 9*X.

To move the graph away from the left edge of the screen in order to allow space for labeling, the horizontal screen position (TS) can be translated toward the right by introducing an additive constant in line number 170 which then becomes: TS = 50 + 50*T. With this change, the graph is translated 50 pixels toward the right side of the screen. After adding line number 90 to provide the set-up for the graphics mode, the program lines shown below will then graph the values calculated using Eq. A.1. The resulting graph is quite similar to the distance-time graph shown in Figure 4.1.

```
 80 REM PROGRAM (A.1)
 90 &HGR2:&MODE(1):&HCOLOR=15:&BCOLOR=0:&PRINT:&MODE(1)
100 V0=10  : REM INITIAL VELOCITY
110 K=5    : REM DRAG COEFFICIENT
120 FOR T=0 TO 10 STEP .1
 130 X=V0*(1-EXP(-K*T))/K : REM CALCULATION OF POSITION AT TIME T
 170 TS=50+50*T    :REM TRANSALATE AND SCALE HORIZONTAL SCREEN LOCATION
```

```
180 XS=90-90*X      :REM TRANSLATE AND SCALE VERTICAL SCREEN
LOCATION
190 &HPLOT TS,XS    :REM PLOT POINT ON SCREEN
200 NEXT T          :REM REPEAT CALCULATION FOR NEXT VALUE OF T
210 END
```

Screen coordinates corresponding to variables x and y are determined by equations of the general form of the statements in line number 170 and line number 180.

$$XS = TX + SX*X \qquad (A.2)$$

$$YS = TY - SY*Y \qquad (A.3)$$

The letter S ending the variable name indicates screen coordinates in the programs of this text. TX and TY specify the translation of the figure and are the screen coordinates of the position corresponding to the point x = 0, y = 0. SX and SY specify the scale of the figure. The values of the scale factors SX and SY are selected to produce figures of the desired size and proportions.

A.6 Graphing a Mathematical Function (Polar Coordinates)

In order to provide an example of the use of polar coordinates, the circle generating program developed below is based on the equation for a circle in polar coordinates: r=constant. The radius of the circle is specified in line number 110. Other polar functions of the form, r=r(θ), can be plotted by altering this line [See, for example, line number 1020 in Program 2.1].

The variable A in the program below represents the angle θ in radians measured with respect to the horizontal (x) axis. The circle is drawn by calculating the cartesian coordinates of the points on the circle in line numbers 120-130 for successive values of θ using the relations x = rcosθ and y = rsinθ. The coordinates of these points are then scaled and translated so that the figure is centered on the screen in line numbers 140-150. (The center of the screen is the location 280,96.) Important! Note that a scale factor of two is used in line number 140 to compensate for the aspect ratio of the pixels of the Beagle Graphics system.

```
80 REM PROGRAM (A.2)
90 &HGR2:&MODE(1):&HCOLOR=15:&BCOLOR=0:&PRINT:&MODE(1)
100 FOR A=0 TO 6.28 STEP .02: REM INCREMENT ANGLE FROM 0 TO
360 DEGREES
110 R=100 : REM RADIUS OF CIRCLE
120 X=R*COS(A) :
130 Y=R*SIN(A)
```

```
140 XS=280+2*X : REM TRANSLATE AND SCALE HORIZONTAL SCREEN
POSITION
150 YS=96-1*Y : REM TRANSLATE AND SCALE VERTICAL SCREEN
POSITION
160 &HPLOT XS,YS
170 NEXT A
180 END
```

A.7 More BASIC Language

For-next loops and graphics statements are the key elements of the programs of this text. Many other useful (but sometimes arcane) operations were avoided in favor of minimizing the need for specialized technical knowledge on the part of users of the programs. With the exception of mathematical operations, the few programming instructions yet to be incorporated in examples are described below. A discussion of the syntax and characteristics of mathematical operations can be found in any BASIC language instruction manual.

The DATA statement is used to store information in the body of a program. In the following example, four numbers are stored within the DATA statement. When called for, the numbers are retrieved sequentially by means of a READ statement.

As the example program below is executed, the stored numbers are printed on the screen in the order in which they occur within the DATA statement. A further example can be found in Program 2.4 where DATA statements are used to store the locations of the vertices of the coordinate system.

Program 2.4 also provides an example of the use of the GOSUB statement. After this statement is encountered, the execution of the program continues at the line number specified in the GOSUB statement. The procedure specified by the GOSUB statement (called a subroutine) is terminated by a RETURN statement. Following the RETURN statement the computer resumes execution of the program at the program line following the original GOSUB statement.

The program listed below provides examples of the DATA, READ, GOSUB, and RETURN statements. The computer reads and then prints the numbers stored in the DATA statement onto the computer screen in the order in which the numbers are found within the DATA statement. Students who are unfamiliar with these statements should experient with the program until the procedures are clearly understood.

```
100 For A= 1 to 4
110 READ X
120 GOSUB 1000
130 NEXT A
140 END
200 DATA 2,4.6,9.3,18.67
1000 PRINT X
1010 RETURN
```

A.8 Program Outline

Graphing of mathematical functions is illustrated by Program A.1 and Program A.2. Both of these programs follow a consistent form: a for-next loop is used to incrementally change the value of the independent variable (time in these examples); the value of the dependent variable is calculated; the variables are then scaled and translated to the appropriate size and position relative to the screen; the figure is plotted.

Because the elements of these programs are uniform and predictable, it is worthwhile to establish an outline to facilitate future development of programs. Highly structured programs can be created in BASIC by identifying the elements of the programs and placing each procedure in a separate subroutine. Programs of this type are based on the principles of programming languages such as PASCAL which require this type of structure.

In order to encourage experimentation, the programs of this text are not organized in terms of specific procedures but instead are organized in terms of the following general operations: graphics set-up, screen display, initial conditions and mechanical description of the system, calculations and plotting. All but the most complex procedures are placed in order of execution within the appropriate section of the program.

Scale factors and translation terms are specified in the program lines which scale the coordinates for plotting directly preceding graphics operations. In this way, graphics operations are clearly separated from the numerical evaluation of equations of motion. All of the variables defined as initial conditions are thus variables defining the physical system being studied rather than the graphics operations required to display the results of calculations.

It is suggested that students establish an outline program suitable for the computer which they use. Program development can then begin by loading the outline program into the computer memory. Students can then develop procedures to perform the required operations for each section of the program.

```
90 REM PROGRAM OUTLINE
100 REM            ***** SET UP GRAPHICS CHARACTERISTICS******

110 & HGR2 : & HCOLOR=15: & BCOLOR=0: & PRINT : & MODE (1)
300 REM
                    *****SET UP SCREEN DISPLAY *******

500 REM
                    *****SPECIFY INITIAL CONDITIONS*****

1000 REM
                  ·***** CALCULATIONS AND PLOTTING *****
```

A.9 Graphics Style Manual

As with other forms of expression, the use of standard style can add to the clarity and acceptability of graphs and drawings. The synopsis of style outlined below paraphrases the style manual of the American Institute of Physics. (See *Style Manual*, American Institute of Physics, 335 East 45th Street, New York, NY 10017.)

1. Graphs should be self-explanatory. Indicate clearly what is being plotted and include units. Keys to symbols should be placed on the graph or in the caption.

2. Do not use capital letters indiscriminately. Use lower case letters for labels. Preserve standard forms for symbols and abbreviations. The vertical axis (ordinate) should be labelled with letters facing upward and printed parallel to the axis. If your graphics system does not have this capability, labels printed horizontally near the top of the axis are usually acceptable.

3. Units should be spaced from the axis labels and enclosed in parentheses. Whenever possible, use the standard prefixes for SI units instead of powers of 10 in axis labels.

4. If powers of 10 are used in axis labels, indicate the power of 10 with the units. For example: X $(10^7 m)$.

5. Use ticks to indicate coordinate values. Ticks should be placed on all four sides of graphs in order to facilitate accurate reading of values from the figure.

6. Graphs should not contain large blank areas. Adjust scale factors so that the range of coordinates does not greatly exceed the range of the graph.

Exercises

A.1. Add labelled vertical and horizontal axes to the graph drawn by Program A.1. See line number 310 to line number 320 of Program 4.1.

A.2. Modify Program A.1 to plot the function y = sin(x) for values of x from 0 to 12.56 radians. Scale the graph so that the amplitude of the function is 50 pixels.

A.3. Modify the program developed in Exercise 2 to mark the horizontal axis with labelled vertical ticks placed every 1.57 radians. See line number 330 to line number 350 in Program 4.4.

A.4. Label the horizontal axis of the figure developed in Exercise 3 "angle (radians)."

A.5. a. Create coordinate axes centered on the screen.
b. Create labelled ticks along the horizontal axis created above from -10 to +10. See line number 330 to line number 355 of Program 5.3.
c. Create labelled ticks from -10 to +10 along the vertical axis of the coordinate system. See line number 357 to line number 380 of Program 5.3. d. Using the labelled axis plot the function $y = [(x^2)/5] - 10$ for values of x from -10 to +10.

A.6. Graph the function $z = -(y^2 + 1)/(y^4 + 8)$ for values of y between -5 and +5. (See Marion and Thornton, Section 2, Example 2.13.)

A.7. Using a semi-log scale, plot a graph of the function $y = 10x$ for values of x between 0 and 4. Place 10 ticks spaced logarithmically on the vertical (y) axis for each order of magnitude. Hint: Calculate y but plot log(y) [.43429*ln(y)]. Don't forget that the BASIC language statement LOG(X) calculates the natural logarithm of x instead of the base 10 logarithm. The resulting graph should be a straight line for this particular function.

A.8. Using a semi-log scale, plot a graph of the function $y = 1/\sin^4(\theta/2)$ for values of θ between 1 degree and 180 degrees. The value of y will vary over nine orders of magnitude. [See Marion and Thornton, Section 8.10.]

A.9. Plot a circle centered at position 200,50.

A.10. a. Alter Program A.1 to draw concentric circles centered on screen.
b. In order to draw concentric equilateral polygons, modify the graphics statement in Program A.2 to draw a line and change step to 6.28/n where n is an integer. What value of n is required to produce a square? What happens if n is not an integer?
c. Change the range of the for-next loop to 20*6.28 and the value of n to 3.95. The concentric circles become a spiraling "string" figure when this change is made.

A.11. a. Modify the circle-drawing program developed above to draw an ellipse by altering the scale factors (line numbers 140-150).
b. Modify the program to draw a family of six concentric ellipses all having the same vertical dimension but with successive ellipses having horizontal dimensions reduced by a constant fraction.

EXERCISES

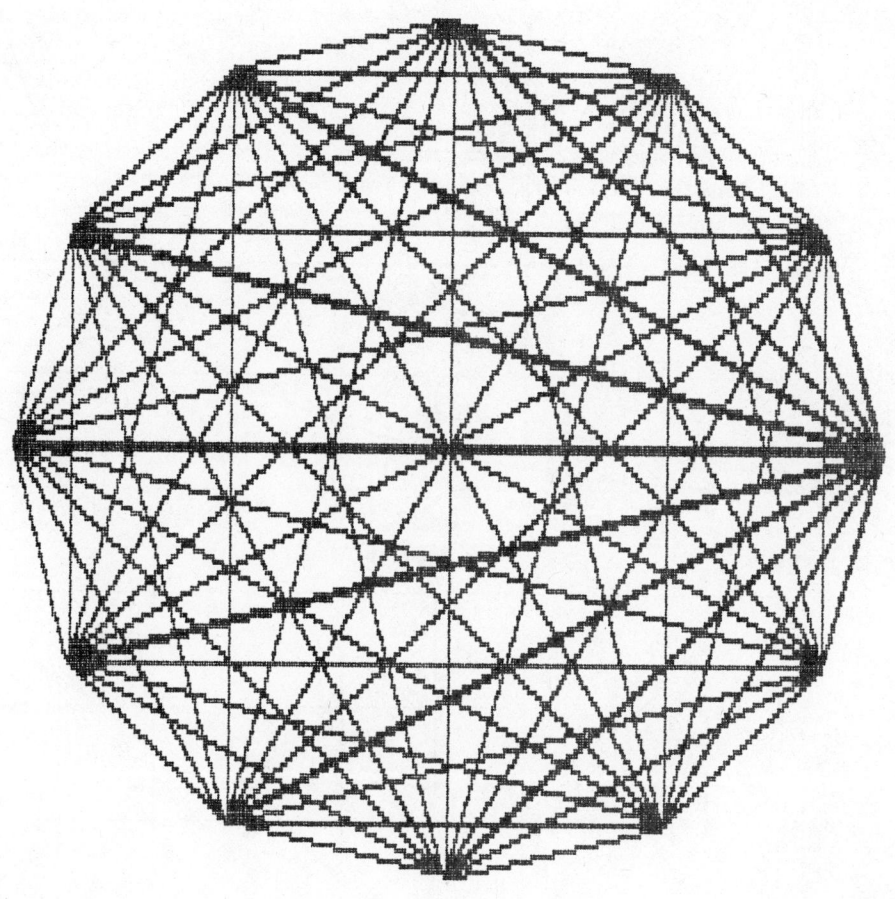

Figure (A.1)

A.12. Develop a program based on the circle-drawing program to draw a figure analogous to Figure A.1 but with any number of vertices specified in the program.

A.13. Structure the programs developed in Exercise A.5 and in Exercise A.8 to fit the suggested program outline.

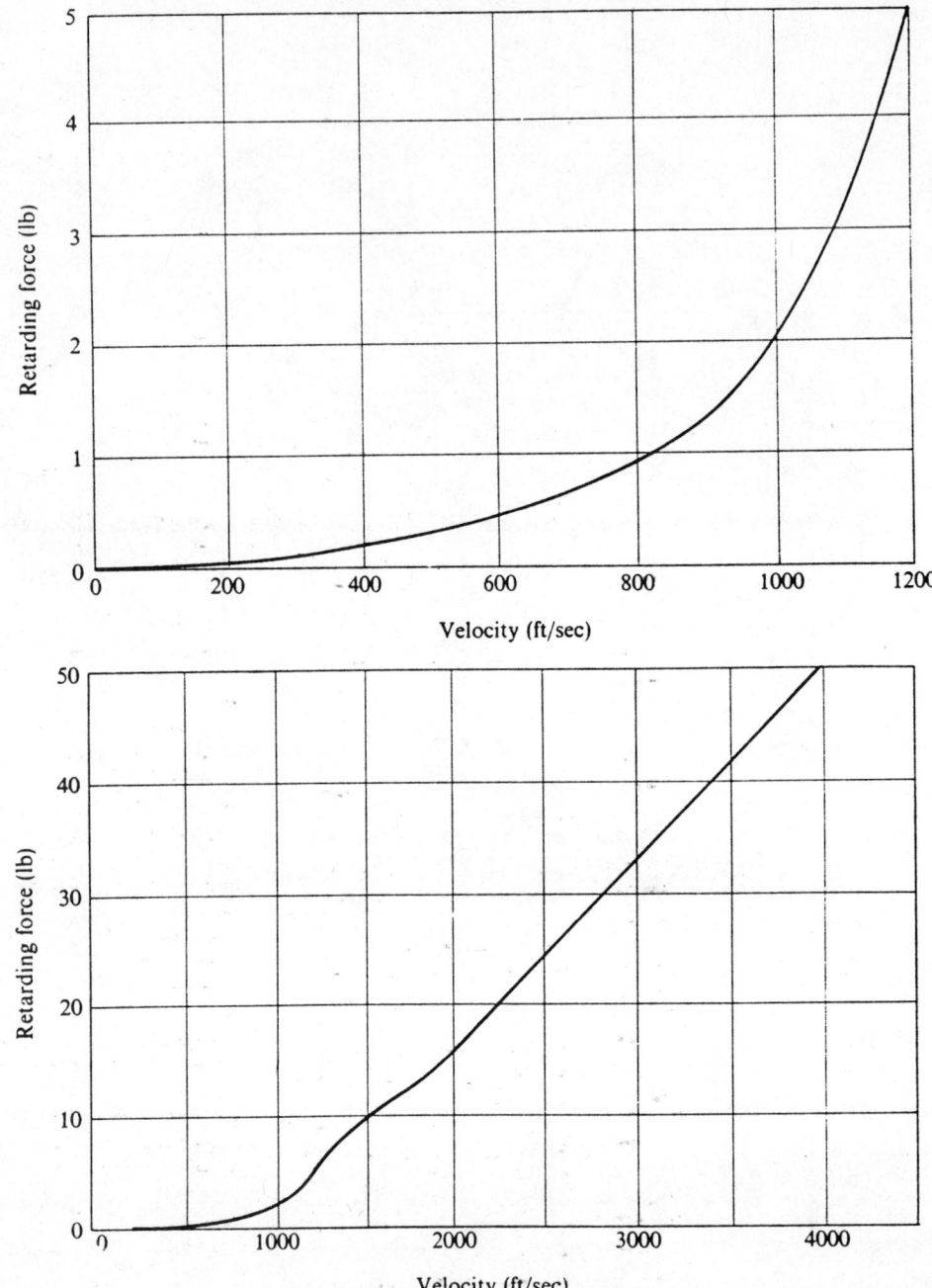

FIGURE A.2

A.14. Figure A.2 shows the retarding force (lb) as a function of velocity (ft/sec) for a particular type of projectile. (See Marion and Thornton, Section 2, Figure 2.3.)

a. Using the top graph, store the value of force associated with the velocity in increments of 100 ft/sec. (If programming in BASIC, an array [F(V/100)] can be conveniently established to store the information for easy access.) Write a program to convert the values to SI units and plot a graph of the values by using the array that was created.

b. Using the bottom graph, store the values of force associated with the velocity in increments of 500 ft/sec. Write a program to convert the values to SI units and graph the values.

Appendix B

Summary of Graphics Statements

B.1 Introduction

In order to create programs suitable for use with a wide range of computers, the programs in this text were written in BASIC using only the standardized elements of the language. For the same reason, programs make use of only four graphics operations: point-plotting, line drawing, cursor positioning, and text printing. Almost all microcomputers currently manufactured are capable of performing this limited set of graphics operations in the context of the BASIC programming language.

This appendix provides detailed information for users of Beagle Graphics for the Apple IIe, GW-BASIC for the IBM-pc, and the DEC ReGIS graphics system for graphics terminals and microcomputers manufactured by Digital Equipment Corporation. Users of other computers can infer how to perform the required graphics operations with their computers even though detailed information is not provided here.

The programs in this text are written for use with Beagle Graphics. To use these programs with any of the systems described in this appendix, the graphics statements must be altered to suit the system which is being used. In most cases, the scale factors must also be altered because of differences in screen resolution and aspect ratios. Appendix C contains listings of the programs for use with the IBM-pc.

B.2 Beagle Graphics

Beagle Graphics is a low cost software package available from Beagle Brothers Software, Inc. or from independant software distributors. The programs allow the Apple IIe and the Apple IIc to perform the desired graphics operations within Applesoft BASIC programs using the Apple's double high resolution graphics display. The program disk contains complete versions of the programs for use with either the DOS 3.3 or the PRODOS operating system.

High resolution graphics was a feature of the original Apple computer and has been included as a standard feature in all of the subsequent versions of the Apple II. The standard high resolution graphics is too limited to be completely satisfactory for use with this text. The most important missing ingredient is the ability to print text on the graphics screen. Without this feature, graphs and drawings cannot be labelled.

Beagle Graphics overcomes this objection by allowing text to be printed on the graphics screen of Apple IIe and Apple IIc computers in addition to doubling the resolution to 560 x 192 pixels. The programs, however, require an Apple IIe computer with at least 128k of RAM. (An Apple extended 80 column card with jumper installed or an equivalent extended memory card is required to provide this memory.) The programs can be used with any model of the Apple IIc but some early models of the Apple IIe must be upgraded to run the programs. The program cannot be used with the Apple II+ computer.

The Beagle Graphics program named DHGR (for double high resolution graphics) must first be installed into the computer memory. In order to print text on the graphics screen, a character set (font) must be loaded after DHGR is installed. In addition, the memory location of the character set must be placed into the appropriate location in the DHGR program. Execution of the program below loads DHGR from a disk and then installs the appropriate character set for use with the programs in this text.

```
10 PRINT CHR$(4);"BRUN DHGR":REM LOAD DHGR INTO MEMORY
20 PRINT CHR$(4);"BLOAD ASCII.FONT,A";16384:REM LOAD FONT STARTING AT MEMORY LOCATION 16384
30 L=PEEK(974)+PEEK(975)*256: REM FIND DHGR LOCATION
40 POKE L+3,0:POKE L+4,64:REM PUT STARTING ADDRESS OF FONT INTO THIRD AND FOURTH BYTE OF DHGR
```

After this program is executed, the command NEW will delete the program from memory but will not affect DHGR or the stored character set. The installed graphics programs do not affect normal operations of the computer. All subsequent programs should include the program line shown below when it is desired to select the double high resolution monochrome graphics display and to enable printing of text on the graphics screen. The command &TEXT returns the computer screen to the normal text mode.

```
100 &HGR2: &HCOLOR=15: &BCOLOR=0: &PRINT: &MODE(1)
```

The graphics operations required for the programs in this text are described in Section A.3. The syntax for each of these operations when using Beagle Graphics is described below. Pixels have an aspect ratio of two in this system.

&HPLOT X,Y

Point-Plotting: This instruction plots a point at the location X,Y. The values of X and Y need not be integers but must correspond to a point on the graphics screen. (See section A.1.2.)

&HPLOT X1,Y1 TO X,Y

Line-Drawing: This instruction draws a line from the screen location X1,Y1 to the screen location X,Y using double high-resolution graphics.

&GOTO X,Y

Cursor-positioning: This instruction positions the graphics cursor at the screen location X,Y. The command is used to position the cursor before text is printed on the graphics screen.

PRINT "Text or variables to be printed"

Text printing: After positioning the graphics cursor, this statement operates as a normal PRINT statement in BASIC. The upper left corner of the first letter of text is positioned at the current location of the graphics cursor.

B.3 GW-BASIC

GW-BASIC is a version of the BASIC programming language developed by Microsoft, Inc. for use with computers that are compatible with the IBM-pc. Specialized versions of GW-BASIC are available for each of the many types of "IBM-pc compatible" computers which use the MS-DOS operating system. Any computer of this type can be used to run the programs in this text. For the IBM Personal Computer in particular, the programs can be used with IBM Personal Computer BASIC

The programming language includes the graphics operations needed for the programs of this text. An appropriate graphics board must be installed in order for the computer to perform these operations.

In order to select the high resolution graphics mode and clear the graphics screen, programs should include the following program line.

```
100 SCREEN 2:CLS
```

The graphics operations required for the programs of this text are described in section A.3. The syntax for each of these operations in GW-BASIC is described below. By replacing only the Beagle Graphics commands in the programs of the text, the programs can be used with the high resolution graphics mode of GW-BASIC. Graphics resolution is 640 x 200 pixels with an aspect ratio of 2.25 in this mode.

PSET(X,Y)

Point-plotting: This instruction plots a point at the screen location X,Y. For the programs of this text, this instruction is equivalent to the Beagle Graphics instruction &HPLOT X,Y.

LINE (X1,Y1)-(X,Y)

Line drawing: This instruction draws a line from the screen location X1,Y1 to the location X,Y. The instruction is equivalent to the Beagle Graphics instruction &HPLOT X1,Y1 to X,Y.

LOCATE Y,X

Cursor-positioning: This instruction positions the text cursor at the location X,Y before text is printed on the screen. It is important to note that when used in this statement, X and Y refer to screen coordinates for the text screen rather than coordinates for the graphics screen. When used with the high resolution graphics screen, the value of X must be in the range from 1 to 80. This number specifies the horizontal location (column) relative to the text screen. The value of Y must be in the range 1 to 25. and indicates the vertical position (row) relative to the text screen. It is also important to notice that the order of X and Y in the LOCATE statement reverses the usual order in graphics statements.

The LOCATE Y,X statement is similar but not equivalent to the Beagle Graphics statement &GOTO X,Y. Because the LOCATE statement controls the text cursor rather than the graphics cursor, text locations and graphics locations require different scale factors. Text is positioned relative to a very low resolution screen with a resolution of 80 x 25 positions and an aspect ratio of 2.25. All scale factors used for graphics must be divided by a factor of 8 when used with the LOCATE statement to determine the screen position for text.

PRINT "Text or variables to be printed"

Text printing: After positioning the text cursor, this statement is used in exactly the same fashion as the familiar PRINT statement in BASIC. The first letter of text is positioned at the current location of the text cursor.

B.4 DEC ReGIS

The DEC ReGIS graphics language is used with terminals and microcomputers manufactured by Digital Equipment Corporation (DEC). The system uses specially equipped terminals and microcomputers (such as the VT-240 and VK-100 terminals, or Rainbow microcomputers) as interpreters for graphics commands imbedded in output from host programs. One particular advantage of this system is the ability of the graphics interpreter to respond to output from almost any computer in the context of almost any programming language.

B.4 DEC REGIS

In the case of BASIC, the graphics interpreter executes graphics commands sent to the interpreter in PRINT statements. Programs must first include a statement (an escape sequence, CHR$(27)+"Pp") to enable the graphics mode. All subsequent PRINT statements will be interpreted as if they are graphics operations rather than text operations. However, the action of other program statements is unaffected. The graphics mode is terminated by the escape character (CHR$(27)). After the graphics mode is terminated, the interpreter treats PRINT statements as normal text operations.

In order to enable the graphics mode and clear the screen, programs in BASIC should begin with the following program line when using DEC ReGIS.

100 PRINT CHR$(27)+"Pp":PRINT"S(E)"

The graphics operations required for the programs of this text are described in Section A.3. The syntax and examples for each of these operations is described below. By replacing only the Beagle Graphics statements in the programs of the text, the programs can be used with computers equipped with interpreters for the DEC ReGIS graphics language. Graphics resolution is 800 x 480 pixels with an aspect ratio of one for most computers and terminals using this system.

PRINT" P[";X;",";Y;"]V[] "

Point-Plotting: This instruction plots a point at the location X,Y. The instruction is equivalent to the Beagle Graphics instruction &HPLOT X,Y. Notice that the syntax for a variable X is ";X;" in this system. The values of variables need not be integers and can even correspond to a location off the graphics screen.

PRINT" P[";X1;",";Y1;"]V[";X;",";Y;"] "

Line Drawing: This instruction draws a line from the screen location X1,Y1 to the screen location X,Y. The instruction is equivalent to the Beagle Graphics instruction &HPLOT X1,Y1 to X,Y.

PRINT" P[";X;",";Y;"] "

Cursor-Positioning: This instruction positions the graphics cursor at the screen location X,Y. The instruction is used to position the graphics cursor before text is printed on the screen. The instruction is also incorporated in the line-drawing operation and in the point-plotting operation. The instruction is equivalent to the Beagle Graphics instruction &GOTO X,Y.

PRINT" T(S1)'Text or variables to be printed' "

Text Printing: After positioning the graphics cursor, this instruction operates in a manner similar to that of a print statement in BASIC. However, the text to be printed must be surrounded by single quotation marks. (For example: T(S1)'distance (m)'.)

The upper left corner of the first letter of text is placed at the current location of the graphics cursor. Text or numbers to be printed on the screen can be incorporated as string variables or numerical variables respectively. (For example: T(S1)' ";A;" ' will print the value of the numerical varible A at the current position of the graphics cursor.) This instruction is equivalent to the PRINT instruction of Beagle Graphics.

A version of Program (2.1) for use with DEC ReGIS is listed below.

```
90   REM : PROGRAM (2.1)
100  REM     ***** SET UP GRAPHICS CHARACTERISTICS *****

110  PRINT CHR$(27)+"Pp":PRINT"S(E)"
300  REM
                ***** SET UP SCREEN DISPLAY *****

310  PRINT"P[1,240]V[760,240]"
320  PRINT"P[380,1]V[380,470]"
330  PRINT"P[600,20]T(S1)'R=R0*sin(3*A)'"
500  REM
                        ***** SPECIFY INITIAL CONDITIONS *****

510  R0 = 90
1000 REM
                        ***** PLOT POLAR FUNCTION *****

1010 FOR A = 0 TO 3.14 STEP .01
1020 R = R0 *  SIN (3 * A)
1030 X = R *  COS (A)
1040 Y = R *  SIN (A)
1070 XS = 280 + 2 * X
1080 YS = 96 - Y
1090 PRINT"P[";XS;",";YS;"]V[]"
1110 NEXT A
2000 END
```

Appendix C

The Fourth Order Runge-Kutta Method

C.1 Introduction

A large number of numerical methods are available for solving ordinary differential equations. Only the simplest methods were considered in the body of this text. Accurate and stable solutions can be developed using these methods to solve the types of problems encountered in the study of classical mechanics.

More complex techniques are useful if highly accurate numerical results are required. For many purposes, the fourth order Runge-Kutta method is chosen as a good compromise combining high accuracy with simplicity. This technique is one of a class of "self-starting" methods that requires the knowledge of only the initial conditions in order to begin the successive calculations leading to a numerical solution of a given differential equation. The purpose of this appendix is to provide a "recipe" for using the fourth order Runge-Kutta method to improve the accuracy of the programs of this text.

C.2 Applying the Fourth Order Runge-Kutta Method

The fourth order Runge-Kutta method is used to provide numerical solutions for first order differential equations. In order to provide an example, Newton's Second Law can be written:

$$dv/dt = F(v,t)/m \qquad (C.1)$$

APPENDIX C

If Euler's method is applied as in Chapter 3, we have:

$$v)_{t+dt} = v)_t + F(v)_t,t)/m \qquad (C.2)$$

If instead the fourth order Runge-Kutta method is selected, calculations proceed as shown below.

$$L_1 = dt[F(v)_t,t)/m] \qquad (C.3)$$

$$L_2 = dt[F(v)_t + L_1/2, t + dt/2)/m] \qquad (C.4)$$

$$L_3 = dt[F(v)_t + L_2/2, t + dt/2)/m] \qquad (C.5)$$

$$L_4 = dt[F(v)_t + L_3, t + dt)/m] \qquad (C.6)$$

$$v)_{t+dt} = v)_t + (L_1 + 2L_2 + 2L_3 + L_4)/6 \qquad (C.7)$$

With the fourth order Runge-Kutta method, four "estimates" of the change of velocity (L_1, L_2, L_3, L_4) are made. The change in velocity for the time increment dt is then determined in Eq. C.7 by calculating a weighted average of the estimates.

Because the technique described above can only be applied to first order differential equations, second order differential equations must be treated as a set of coupled first order differential equations. Again using Newton's Second Law as an example, we have:

$$dv/dt = F(x,v,t)/m \qquad (C.8)$$

$$dx/dt = v \qquad (C.9)$$

Applying the fourth order Runge-Kutta method to this set of coupled equations, calculations proceed as shown below.

$$K_1 = dt(v)_t) \qquad (C.10)$$

$$L_1 = dt[F(x)_t,v)_t,t)/m] \qquad (C.11)$$

$$K_2 = dt(v)_t + L_1/2) \qquad (C.12)$$

$$L_2 = dt[F(x)_t + K_1/2, v)_t + L_1/2, t + dt/2)/m] \qquad (C.13)$$

$$K_3 = dt(v)_t + L_2/2) \qquad (C.14)$$

$$L_3 = dt[F(x)_t + K_2/2, v)_t + L_2/2, t + dt/2)/m] \qquad (C.15)$$

$$K_4 = dt(v)_t + L_3) \qquad (C.16)$$

$$L_4 = dt[F(x)_t + K_3, v)_t + L_3, t + dt)/m] \qquad (C.17)$$

C.2 APPLYING THE FOURTH ORDER RUNGE-KUTTA METHOD

$$x)_{t+dt} = x)_t + (K_1 + 2K_2 + 2K_3 + K_4)/6 \qquad (C.18)$$

$$v)_{t+dt} = v)_t + (L_1 + 2L_2 + 2L_3 + L_4)/6 \qquad (C.19)$$

By starting with the initial values of position (x) and velocity (v) at t=0, subsequent values of x and v can be determined.

For other differential equations, the method is similarly applied. Dependent variables replace the position (x) and velocity (v) while the time (t) is replaced by the independent variable. Coupled sets of second order differential equations can be evaluated numerically by extending the system of coefficients in a method analogous to that described above.

Program 4.1 can be converted from Euler's method to the fourth order Runge-Kutta method by first deleting line number 1020 and line number 1030 and then inserting the following lines into the program.

```
1012 K1=DT*(-K*V):REM Eq. C.10
1014 L1=DT*V:REM Eq. C.11
1016 K2=DT*(-K*(V+(K1/2))):REM Eq. C.12
1018 L2=DT*(V+(K1/2)):REM Eq. C.13
1020 K3=DT*(-K*(V+(K2/2))):REM Eq. C.14
1022 L3=DT*(V+(K2/2)):REM Eq. C.15
1024 K4=DT*(-K*(V+K3)):REM Eq. C.16
1026 L4=DT*(V+K3):REM Eq. C.17
1030 V1=V+(K1+2*K2+2*K3+K4)/6:REM Eq. C.19
1034 X1=X+(L1+2*L2+2*L3+L4)/6:REM Eq. C.18
```

APPENDIX D

Program Listings for the IBM-pc

D.1 Program (2.1)

```
90   REM : PROGRAM (2.1)
100  REM             ***** SET UP GRAPHICS CHARACTERISTICS *****

110  SCREEN 2:CLS
300  REM
                     *****  SET UP SCREEN DISPLAY *****

310  LINE(1,96)-(630,96)
320  LINE(320,1)-(320,190)
330  LOCATE 2,55:PRINT"R=R0*sin(3*A)"
500  REM
                     ***** SPECIFY INITIAL CONDITIONS *****

510  R0 = 90
1000 REM
                     *****   PLOT POLAR FUNCTION *****

1010 FOR A = 0 TO 3.14 STEP .01
1020 R = R0 *  SIN (3 * A)
1030 X = R *  COS (A)
1040 Y = R *  SIN (A)
1070 XS = 320 + 2.25 * X:REM GRAPHICS SCREEN HAS AN ASPECT RATIO OF
2.25
1080 YS = 96 - Y
1090  PSET(XS,YS)
1110  NEXT A
2000  END
```

D.2 Program (2.2)

```
90   REM : PROGRAM (2.2)
100  REM             ***** SET UP GRAPHICS CHARACTERISTICS *****

110  SCREEN 2:CLS
```

```
300  REM
                    *****  SET UP SCREEN DISPLAY *****

310  LINE(1,96)-(630,96)
320  LINE(320,1)-(320,190)
500  REM
                 *****  SPECIFY INITIAL CONDITIONS *****

510  R0 = 90
520  A1 = .3: REM  ROTATION ANGLE
1000 REM
                    *****  PLOT POLAR FUNCTION *****

1010  FOR A = 0 TO 3.14 STEP .01
1020  R = R0 *  SIN (3 * A)
1030  X = R *  COS (A)
1040  Y = R *  SIN (A)
1050  X1 = X *  COS (A1) + Y *  SIN (A1)
1060  Y1 =  - X *  SIN (A1) + Y *  COS (A1)
1070  XS = 320 + 2.25 * X1
1080  YS = 96 - Y1
1090   PSET(XS,YS)
1110   NEXT A
2000  END
```

D.3 Program (2.3)

```
90  REM  PROGRAM (2.3)
100 REM          *****  SET UP GRAPHICS CHARACTERISTICS *****

110  SCREEN 2:CLS
300  REM
                    *****  SET UP SCREEN DISPLAY *****

310  LINE(1,96)-(630,96)
320  LINE(320,1)-(320,190)
500  REM
                 *****  SPECIFY INITIAL CONDITIONS *****

510  A = 90
520  B = 30
1000  REM
              *****  CALCULATE VALUES AND PLOT FUNCTION *****
```

```
1010  FOR TH = 0 TO 6.28 STEP .02
1020  X = A *  COS (TH)
1030  Y = B *  SIN (TH)
1040  XS = 320 + 2.25 * X
1050  YS = 96 - Y
1060  PSET(XS,YS)
1070  NEXT TH
2000  END
```

D.4 Program (2.4)

```
7    REM    PROGRAM (2.4)
10   REM    THIS PROGRAM USES EULER'S ANGLES FOR ROTATION
80   REM
```
***** SET UP GRAPHICS CHARACTERISTICS *****

```
90   SCREEN 2:CLS
95   REM
```
***** SET UP SCREEN DISPLAY *****

```
100  REM      ANGLE A1 IS PHI:REM ANGLE A2 IS THETA:REM ANGLE A3 IS
PSI:REM ALL ANGLES ARE EXPRESSED IN RADIANS
110  A1 = 0
120  A2 = 0
130  A3 = 0
135  LOCATE 1,65: PRINT "A1=";A1
137  LOCATE 2,65: PRINT "A2=";A2
139  LOCATE 3,65: PRINT "A3=";A3
140  S1 =  SIN (A1):C1 =  COS (A1)
150  S2 =  SIN (A2):C2 =  COS (A2)
160  S3 =  SIN (A3):C3 =  COS (A3)
170  REM                                                      L1  L2  L3
180  REM  CALCULATE ELEMENTS OF TRANSFORMATION MATRIX          L4  L5  L6
190  REM                                                      L7  L8  L9
200  L1 = C3 * C1 - C2 * S1 * S3
210  L2 =  - S3 * C1 - C2 * S1 * C3
220  L3 = S2 * S1
230  L4 = C3 * S1 + C2 * C1 * S3
240  L5 =  - S3 * S1 + C2 * C1 * C3
250  L6 =  - S2 * C1
260  L7 = S3 * S2
270  L8 = C3 * S2
280  L9 = C2
```

```
290 D = 20: REM              DISTANCE OF VIEWER FROM SCREEN (CM)
300 XV = D * L1
310 YV = D * L2
320 ZV = D * L3
330  READ A,X1,Y1,Z1,X,Y,Z
340  IF A = 16 GOTO 5000
350  GOSUB 2000
490 SX = XS:SZ = ZS:X = X1:Y = Y1:Z = Z1
495  GOSUB 2000
500  LINE(XS,ZS)-(SX,SZ)
510  IF A = 4 THEN   LOCATE ZS/8,XS/8: PRINT "X"
520  IF A = 11 THEN  LOCATE ZS/8,XS/8: PRINT "Y"
530  IF A = 15 THEN  LOCATE ZS/8,XS/8: PRINT "Z"
540  GOTO 330
1900 REM
                    ***** ROTATION SUBROUTINE *****

2000 XO = X - XV: REM       TRANSLATION OF COORDINATES TO MOVE
OBSERVER TO ORIGIN OF COORDINATES
2010 YO = Y - YV
2020 ZO = Z - ZV
2025 REM       ***** APPLY ROTATION MATRIX ****
2030 X3 = L1 * XO + L2 * YO + L3 * ZO
2040 Y3 = L4 * XO + L5 * YO + L6 * ZO
2050 Z3 = L7 * XO + L8 * YO + L9 * ZO
2060 REM       ***** PROJECT ROTATED OBJECT ON SCREEN *****
2070 XS = 320 + 45 * D * Y3 / ( - X3)
2080 ZS = 96 - 20 * D * Z3 / ( - X3)
2090 RETURN
3000 REM
                    ***** DATA FOR SCREEN DISPLAY *****

3010 DATA    1,3,3,0,-3,3,0
3020 DATA    2,3,2,0,-3,2,0
3030 DATA    3,3,1,0,-3,1,0
3040 DATA    4,4,0,0,-4,0,0
3050 DATA    5,3,-1,0,-3,-1,0
3060 DATA    6,3,-2,0,-3,-2,0
3070 DATA    7,3,-3,0,-3,-3,0
3080 DATA    8,3,3,0,3,-3,0
3090 DATA    9,2,3,0,2,-3,0
3100 DATA    10,1,3,0,1,-3,0
3110 DATA    11,0,4,0,0,-4,0
3120 DATA    12, -1,3,0,-1,-3,0
3130 DATA    13,-2,3,0,-2,-3,0
```

```
3140   DATA    14,-3,3,0,-3,-3,0
3150   DATA    15,0,0,4,0,0,0
3160   DATA    16,0,0,0,0,0,0
5000   END
```

D.5 Program (2.5)

```
7   REM     PROGRAM (2.5)
10  REM     THIS PROGRAM USES EULER'S ANGLES FOR ROTATION
80  REM
                     ***** SET UP GRAPHICS CHARACTERISTICS *****

90  SCREEN 2:CLS
95  REM
                     ***** SET UP SCREEN DISPLAY *****

100 REM         ANGLE A1 IS PHI:REM ANGLE A2 IS THETA:REM ANGLE A3 IS
PSI:REM ALL ANGLES ARE EXPRESSED IN RADIANS
110 A1 = 0
120 A2 = 0
130 A3 = 0
135   LOCATE  1,65: PRINT "A1=";A1
137   LOCATE  2,65: PRINT "A2=";A2
139   LOCATE  3,65: PRINT "A3=";A3
140 S1 =   SIN (A1):C1 =   COS (A1)
150 S2 =   SIN (A2):C2 =   COS (A2)
160 S3 =   SIN (A3):C3 =   COS (A3)
170 REM                                                   L1  L2  L3
180 REM    CALCULATE ELEMENTS OF TRANSFORMATION MATRIX    L4  L5  L6
190 REM                                                   L7  L8  L9
200 L1 = C3 * C1 - C2 * S1 * S3
210 L2 =   - S3 * C1 - C2 * S1 * C3
220 L3 = S2 * S1
230 L4 = C3 * S1 + C2 * C1 * S3
240 L5 =   - S3 * S1 + C2 * C1 * C3
250 L6 =   - S2 * C1
260 L7 = S3 * S2
270 L8 = C3 * S2
280 L9 = C2
290 D = 20: REM          DISTANCE OF VIEWER FROM SCREEN (CM)
300 XV = D * L1
310 YV = D * L2
320 ZV = D * L3
330   READ A,X1,Y1,Z1,X,Y,Z
```

```
340   IF A = 16 GOTO 1000
350   GOSUB 2000
490 SX = XS:SZ = ZS:X = X1:Y = Y1:Z = Z1
495   GOSUB 2000
500   LINE(XS,ZS)-(SX,SZ)
510   IF A = 4 THEN    LOCATE  ZS/8,XS/8: PRINT "X"
520   IF A = 11 THEN   LOCATE  ZS/8,XS/8: PRINT "Y"
530   IF A = 15 THEN   LOCATE  ZS/8,XS/8: PRINT "Z"
540   GOTO 330
1000  REM
                ***** CALCULATE VALUES AND PLOT FUNCTION *****

1010  FOR PH = 0 TO 3.14 STEP .01
1020 X = 3 *  COS (3 * PH) *  COS (PH)
1030 Y = 3 *  COS (3 * PH) *  SIN (PH)
1040 Z = 3 * ( COS (3 * PH) ^ 2)
1050  GOSUB 2000
1060  PSET(XS,ZS)
1070  NEXT PH
1080  GOTO 5000
1900  REM
                    ***** ROTATION SUBROUTINE *****


2000 XO = X - XV: REM        TRANSLATION OF COORDINATES TO MOVE
OBSERVER TO ORIGIN OF COORDINATES
2010 YO = Y - YV
2020 ZO = Z - ZV
2025  REM     ***** APPLY ROTATION MATRIX ****
2030 X3 = L1 * XO + L2 * YO + L3 * ZO
2040 Y3 = L4 * XO + L5 * YO + L6 * ZO
2050 Z3 = L7 * XO + L8 * YO + L9 * ZO
2060  REM     ***** PROJECT ROTATED OBJECT ON SCREEN *****
2070 XS = 320 + 45 * D * Y3 / ( - X3)
2080 ZS = 96 - 20 * D * Z3 / ( - X3)
2090  RETURN
3000  REM
                  ***** DATA FOR SCREEN DISPLAY *****

3010  DATA   1,3,3,0,-3,3,0
3020  DATA   2,3,2,0,-3,2,0
3030  DATA   3,3,1,0,-3,1,0
3040  DATA   4,4,0,0,-4,0,0
3050  DATA   5,3,-1,0,-3,-1,0
3060  DATA   6,3,-2,0,-3,-2,0
```

```
3070  DATA  7,3,-3,0,-3,-3,0
3080  DATA  8,3,3,0,3,-3,0
3090  DATA  9,2,3,0,2,-3,0
3100  DATA  10,1,3,0,1,-3,0
3110  DATA  11,0,4,0,0,-4,0
3120  DATA  12,-1,3,0,-1,-3,0
3130  DATA  13,-2,3,0,-2,-3,0
3140  DATA  14,-3,3,0,-3,-3,0
3150  DATA  15,0,0,4,0,0,0
3160  DATA  16,0,0,0,0,0,0
5000  END
```

D.6 Program (4.1)

```
90   REM   PROGRAM (4.1)
100  REM          ***** SET UP GRAPHICS CHARACTERISTICS *****

110  SCREEN 2:CLS
300  REM
                 *****  SET UP SCREEN DISPLAY *****

310  LINE(50,0)-(50,190): REM     DRAW VERTICAL AXES
320  LINE(50,90)-(630,90): REM      DRAW UPPER HORIZONTAL (TIME) AXIS
330  LINE(50,180)-(630,180): REM     DRAW LOWER HORIZONTAL (TIME) AXIS
340  LOCATE 12,32: PRINT "TIME (s)"
350  LOCATE 23,32: PRINT "TIME (s)"
360  LOCATE 2,1: PRINT "VELOCITY"
370  LOCATE 3,2: PRINT "(m/s)"
380  LOCATE 13,1: PRINT "DISTANCE"
390  LOCATE 14,3: PRINT "(m)"
500  REM
               ***** SPECIFY INITIAL CONDITIONS *****

510  X = 0
520  V = 10
530  K = 2
540  DT = .1
1000 REM
             *****  CALCULATE AND PLOT VELOCITY AND POSITION *****

1010 FOR T = 0 TO 10 STEP DT
1020 V1 = V - K * V * DT: REM    EQ. 4.4
```

```
1030 X1 = X + V * DT: REM       EQ. 4.5
1040 TS = 50 + 50 * T: REM        TS IS  HORIZONTAL SCREEN POSITION TO
PLOT TIME
1050 VS = 90 - 9 * V: REM         VS IS VERTICAL SCREEN POSITION TO
PLOT VELOCITY
1060  PSET(TS,VS)
1070 XS = 180 - 10 * X: REM       XS IS VERTICAL SCREEN POSITION TO
PLOT POSITION
1080  PSET(TS,XS)
1110 X = X1:V = V1
1120  LOCATE 20,50 : PRINT "X=";X;" m   "
1130  LOCATE 3,50: PRINT "V=";V;" m/s    "
1140  NEXT T
2000  END
```

D.7 Program (4.2)

```
90  REM   PROGRAM (4.2)
100  REM         ***** SET UP GRAPHICS CHARACTERISTICS *****

110  SCREEN 2:CLS
300  REM
                     *****   SET UP SCREEN DISPLAY *****

310  LINE(50,0)-(50,160): REM  DRAW VERTICAL (VELOCITY) AXIS
320  LINE(50,160)-(630,160): REM  DRAW HORIZONTAL (TIME) AXIS
330  FOR T = 0 TO 10 STEP 2
340  X = 50 + 50 * T
350  LOCATE 20,X/8: LINE(X,160 - 3)-(X,160): LOCATE 21,X/8: PRINT
T: REM   LABEL HORIZONTAL AXIS
360  NEXT T
370  LOCATE 23,36: PRINT "TIME (s)"
380  FOR V = 0 TO 1500 STEP 200
390  Y = 160 - .1 * V
400  LINE(50,Y)-(50 + 6,Y):LOCATE Y/8,2: PRINT V: REM   LABEL
VERTICAL AXIS
410  NEXT V
420  LOCATE 1,1: PRINT "VELOCITY (m/s)"
500  REM
                    ***** SPECIFY INITIAL CONDITIONS *****

510  V = 0: REM    INITIAL VELOCITY
520  VE = 300: REM    EXHAUST VELOCITY
530  DM = 100: REM    RATE AT WHICH FUEL IS BURNED (DM/DT)
540  MP = 10: REM    MASS OF PAYLOAD
```

```
550  M0 = 1000: REM    INITIAL MASS OF ROCKET
560  T1 = .98 * M0 / DM: REM    TIME TO BURN ALL FUEL
570  DT = .1: REM    TIME INCREMENT BETWEEN CALCULATIONS
1000 REM
                    ***** CALCULATE AND PLOT VELOCITY *****

1010 FOR T = 0 TO T1 STEP DT
1020 V1 = V + (VE * DM / (M0 + MP - DM * T)) * DT: REM    EQ. 4.10
1030 YS = 160 - .1 * V: REM       VERTICAL SCREEN POSITION TO PLOT V
1040 XS = 50 + 50 * T: REM       HORIZONTAL SCREEN POSITION TO PLOT T
1050 PSET(XS,YS)
1060 V = V1:X = X1
1070 NEXT T
2000 END
```

D.8 Program (4.3)

```
90   REM   PROGRAM (4.3)
100  REM           ***** SET UP GRAPHICS CHARACTERISTICS *****

110  SCREEN 2:CLS
300  REM
                    *****  SET UP SCREEN DISPLAY *****

310  LINE(50,0)-(50,160): REM   DRAW VERTICAL (VELOCITY) AXIS
320  LINE(50,160)-(630,160): REM   DRAW HORIZONTAL (TIME) AXIS
330  FOR T = 0 TO 10 STEP 2
340  X = 50 + 50 * T
350  LINE(X,160 - 3)-(X,160):LOCATE 21,X/8: PRINT T: REM   LABEL
HORIZONTAL AXIS
360  NEXT T
370  LOCATE 22,36: PRINT "TIME (s)"
380  FOR V = 0 TO 1500 STEP 200
390  Y = 160 - .1 * V
400  LINE(50,Y)-(50 + 6,Y): LOCATE Y/8,2: PRINT V: REM   LABEL
VERTICAL AXIS
410  NEXT V
420  LOCATE 1,2: PRINT "VELOCITY (m/s)"
500  REM
                    ***** SPECIFY INITIAL CONDITIONS *****

510  V = 0: REM    INITIAL VELOCITY
520  VE = 300: REM    EXHAUST VELOCITY (BOTH STAGES)
530  DM = 100: REM    RATE AT WHICH FUEL IS BURNED (DM/DT FOR BOTH
STAGES)
```

```
540 MP = 10: REM      MASS OF PAYLOAD
550 M10 = 700: REM      INITIAL MASS OF FIRST STAGE
560 M20 = 300: REM   INITIAL MASS OF SECOND STAGE
600 T1 = .95 * M10 / DM: REM      TIME TO BURN ALL FUEL IN FIRST
STAGE
610 T2 = .97 * M20 / DM: REM   TIME TO BURN ALL FUEL IN SECOND STAGE
620 DT = .1: REM      TIME INCREMENT BETWEEN CALCULATIONS
1000 REM
                    *****   CALCULATE AND PLOT VELOCITY *****

1010  FOR T = 0 TO T1 STEP DT
1020  V1 = V + (VE * DM / (M10 + M20 + MP - DM * T)) * DT: REM EQ.
4.10
1030  YS = 160 - .1 * V: REM      VERTICAL SCREEN POSITION TO PLOT V
1040  XS = 50 + 50 * T: REM      HORIZONTAL SCREEN POSITION TO PLOT T
1050  PSET(XS,YS)
1060  V = V1:X = X1
1070  NEXT T
1080  LINE(XS,YS - 3)-(XS,YS + 3): REM   MARKER FOR SECOND STAGE
IGNITION
1090  FOR T = T1 TO T1 + T2 STEP DT
1100  V1 = V + (VE * DM / (M20 + MP - DM * (T - T1))) * DT
1110  YS = 160 - .1 * V
1120  XS = 50 + 50 * T
1130  PSET(XS,YS)
1140  V = V1:X = X1
1150  NEXT T
2000  END
```

D.9 Program (4.4)

```
90  REM  PROGRAM (4.4)
100 REM         *****  SET UP GRAPHICS CHARACTERISTICS *****

110  SCREEN 2:CLS
300  REM
                      *****   SET UP SCREEN DISPLAY *****

310  LINE(50,0)-(50,190): REM   DRAW VERTICAL AXIS
320  LINE(50,96)-(630,96): REM   DRAW HORIZONTAL AXIS
330  FOR T = 1 TO 3
340  XS = 50 + 150 * T: LINE(XS,99)-(XS,93): LOCATE 13,XS/8 : PRINT
T
350  NEXT T
360  LOCATE 14,36: PRINT "TIME (s)"
```

116   APPENDIX D

```
370   FOR A = 3 TO - 3 STEP - 1
380  YS = 96 - 20 * A
390   LINE(50,YS)-(55,YS):LOCATE YS/8,3: PRINT A
400   NEXT A
410   LOCATE 1,1: PRINT "ANGLE"
420   LOCATE 2,1: PRINT "(radians)"
500   REM
                    ***** SPECIFY INITIAL CONDITIONS *****

510 A = 1: REM       INITIAL ANGLE IN RADIANS
520 V = - 4: REM        INITIAL ANGULAR VELOCITY OF PENDULUM IN
RADIANS/SECOND
530 L = 1: REM   LENGTH OF PENDULUM IN METERS
540 G = 9.8: REM     ACCELERATION DUE TO GRAVITY
550 DT = .01
1000   REM
                  *****  CALCULATE AND PLOT ANGULAR  POSITION *****

1010 V = V - (G *  SIN (A) / L) * DT / 2: REM     VELOCITY AT DT/2
1020   FOR T = 0 TO 3 STEP DT
1030 A1 = A + V * DT: REM  EQ. 4.11
1040 V1 = V - (G *  SIN (A1) / L) * DT: REM      EQ. 4.12
1050 XS = 50 + 150 * T
1060 YS = 96 - 20 * A
1070   PSET(XS,YS)
1080 V = V1:A = A1
1090   NEXT T
2000   END
```

D.10   Program (5.1)

```
90   REM   PROGRAM (5.1)
100   REM    ***** SET UP GRAPHICS CHARACTERISTICS *****

110   SCREEN 2:CLS
300   REM
                       *****  SET UP SCREEN DISPLAY *****


500   REM
                       ***** SPECIFY INITIAL CONDITIONS *****

510 V1 = .5: REM   INITIAL X COMPONENT OF VELOCITY
520 V3 = 0: REM   INITIAL Y COMPONENT OF VELOCITY
530 X = .7: REM    INITIAL POSITION RELATIVE TO HORIZONTAL (X) AXIS
540 Y = .7: REM    INITAL POSITION RELATIVE TO VERTICAL (Y) AXIS
```

```
550 KX = 9: REM   SPRING CONSTANT IN X DIRECTION
560 KY = 16: REM   SPRING CONSTANT IN Y DIRECTION
570 M = 1: REM   MASS OF OSCILLATOR
580 DT = .01: REM   TIME INCREMENT
1000  REM
                                        ***** CALCULATE AND PLOT POSITION *****
1010 V1 = V1 - (KX / M) * X * DT / 2: REM   Eq. 5.3
1020 V3 = V3 - (KY / M) * Y * DT / 2: REM   Eq. 5.5
1030  FOR T = 0 TO 100 STEP DT
1040 X1 = X + V1 * DT: REM   Eq. 5.4
1050 Y1 = Y + V3 * DT: REM   Eq. 5.6
1060 V2 = V1 - (KX / M) * X1 * DT
1070 V4 = V3 - (KY / M) * Y1 * DT
1080 XS = 320 + 180 * X
1090 YS = 96 - 80 * Y
1100  PSET(XS,YS)
1110 V1 = V2:V3 = V4
1120 X = X1:Y = Y1
1130  NEXT T
2000  END
```

D.11 Program (5.2)

```
90  REM  PROGRAM(5.2)
100  REM   ***** SET UP GRAPHICS CHARACTERISTICS *****

110  SCREEN 2:CLS
300  REM
                                        ***** SET UP SCREEN DISPLAY *****

500  REM
                                        ***** SPECIFY INITIAL CONDITIONS *****

510 X = 1
520 Y = 0
530 V1 = 0
540 V3 = 0
550 L = 10
560 W = 7.27 * 10 ^ - 2: REM   ANGULAR VELOCITY OF ROTATING SYSTEM
(1000 TIMES THE ANGULAR VELOCUTY OF THE EARTH)
570 G = 9.8
580 DT = .05
1000  REM
         ***** CALCULATE AND PLOT VELOCITY AND POSITION *****
```

```
1010   FOR T = 0 TO 1000 STEP DT
1020   V2 = V1 + ( - G * X / L + 2 * W * V3) * DT
1030   V4 = V3 + ( - G * Y / L - 2 * W * V2) * DT
1040   X1 = X + V2 * DT
1050   Y1 = Y + V4 * DT
1060   XS = 320 + 180 * X
1070   YS = 96 - 80 * Y
1080    PSET(XS,YS)
1090   V1 = V2:V3 = V4
1100   X = X1:Y = Y1
1110   NEXT T
2000   END
```

D.12 Program (5.3)

```
90   REM    PROGRAM (5.3)
100  REM     ***** SET UP GRAPHICS CHARACTERISTICS *****

110  SCREEN 2:CLS
300  REM
                    *****   SET UP SCREEN DISPLAY *****

310   LINE(1,96)-(630,96)
320   LINE(320,1)-(320,191)
330   FOR X = 10 ^ 7 TO 6 * 10 ^ 7 STEP 10 ^ 7
340  XS = 320 + X * 4.5 * 10 ^  - 6: REM   SCALE HORIZONTAL AXIS
350   LINE(XS,96 - 2)-(XS,96 + 2): LOCATE 13,XS/8 : PRINT X / 10 ^ 7
355   NEXT X
357   FOR Y = 10 ^ 7 TO 4 * 10 ^ 7 STEP 10 ^ 7
360  YS = 96 - Y * 2 * 10 ^  - 6: REM   SCALE VERTICAL AXIS
370  LINE(320 - 4,YS)-(320 + 4,YS): LOCATE YS/8,41: PRINT Y / 10 ^ 7
380   NEXT Y
390   LOCATE 14,49: PRINT "R(10^7m)"
410   FOR A = 0 TO 6.28 STEP .04:XC = 320 + 29 *  COS (A):YC = 96 -
13 *  SIN (A): PSET(XC,YC): NEXT A: REM    DRAW CIRCLE AT CENTER OF
SCREEN
500  REM
                    ***** SPECIFY INITIAL CONDITIONS *****

510  R = 5.23 * 10 ^ 7: REM   INITIAL VALUE OF R (DISTANCE FROM EARTH
TO SATELLITE)
515  A = 3: REM  INITIAL ANGULAR POSITION
520  V = 0: REM   INITIAL RADIAL VELOCITY
530  VA = 3.5 * 10 ^  - 5: REM        INITIAL ANGULAR VELOCITY
540  G = 6.67 * 10 ^  - 11: REM   UNIVERSAL GRAVITATIONAL CONSTANT
```

```
550 M = 6 * 10 ^ 24: REM   MASS OF EARTH
560 M1 = 100: REM    MASS OF SATELLITE (REDUCED MASS OF SYSTEM)
570 L = M1 * (R ^ 2) * VA: REM   ANGULAR MOMENTUM OF SYSTEM
580 DT = 500: REM    TIME INCREMENT
1000  REM
                    ***** CALCULATE AND PLOT VELOCITY AND
POSITION *****
1020  FOR T = 0 TO 800000 STEP DT
1025 V1 = V + (((  - G * M) / (R ^ 2)) + ((L ^ 2) / ((M1 ^ 2) * (R ^
3)))) * DT
1030 R1 = R + V1 * DT
1040 A1 = A + (L / (M1 * (R1 ^ 2))) * DT
1060 X = R *  COS (A):Y = R *  SIN (A)
1070 XS = 320 + X * 4.5 * (10 ^  - 6):YS = 96 - Y * 2 * (10 ^  - 6)
1080  PSET(XS,YS)
1090 V = V1:R = R1:A = A1
1100  NEXT T
2000  END
```

D.13 Program (5.4)

```
90  REM   PROGRAM (5.4)
100  REM    ***** SET UP GRAPHICS CHARACTERISTICS *****

110  SCREEN 2:CLS
300  REM
                        *****   SET UP SCREEN DISPLAY *****

310  LINE(320,1)-(320,190)
320  FOR X =  - 9 TO 9
330  XS = 320 + X * 101 / 3: REM    SCALE SCREEN HORIZONTALLY(37
PIXELS=1*10^6M)
340  LINE(XS,96 + 3)-(XS,96 - 3): LOCATE 13,XS/8: PRINT X
350  NEXT X
360  LOCATE 13,44: PRINT "DISTANCE (10^6m)"
500  REM
                    ***** SPECIFY INITIAL CONDITIONS *****

510 MM = 7.4 * 10 ^ 22: REM    MASS OF THE MOON (KG)
520 M = 100: REM   MASS OF SATELLITE (KG)
530 G = 6.67 * 10 ^  - 11: REM    UNIVERSAL GRAVITATIONAL CONSTANT
540 VA = 930: REM       VELOCITY OF MOON (HORIZONTAL MOTION RELATIVE
TO THE SCREEN)
545 VB = 0
```

```
550 VX = 0: REM    INITIAL HORIZONTAL COMPONENT OF VELOCITY OF
SATELLITE (RELATIVE TO SCREEN)
560 VY = 0: REM    INITIAL VERTICAL COMPONENT OF VELOCITY OF
SATELLITE (RELATIVE TO SCREEN)
570 A = - 9 * 10 ^ 6: REM   INITIAL X COORDINATE OF MOON
580 B = 0: REM    INITIAL Y COORDINATE OF MOON
590 X = 0: REM    INITIAL X COORDINATE OF SATELLITE
600 Y = - 3 * 10 ^ 6: REM   INITIAL Y COORDINATE OF SATELLITE
610 DT = 100: REM  TIME INCREMENT
1000  REM
                ***** CALCULATE AND PLOT VELOCITY AND POSITION *****

1010  FOR T = 0 TO 12000 STEP DT
1020 A = A + VA * DT:UM = 320 + A * (101 / (3 * 10 ^ 6)):
PSET(UM,96): REM       PLOT PATH OF MOON
1030 R =  SQR ((A - X) ^ 2 + (B - Y) ^ 2): REM   DISTANCE BETWEEN
MOON AND SATELLITE (M)
1040 F = G * MM * M / (R * R): REM   GRAVITATIONAL FORCE BETWEEN
MOON AND SATELLITE
1050 FX = F * (A - X) / R: REM    X-COMPONENT OF GRAVITATIONAL FORCE
1060 FY = F * (B - Y) / R: REM    Y-COMPONENT OF GRAVITATIONAL FORCE
1070 VX = VX + (FX / M) * DT
1080 VY = VY + (FY / M) * DT
1090 X1 = X + VX * DT: REM  LAST POINT APPROXIMATION
1100 Y1 = Y + VY * DT
1110 XS = 320 + X * (101 / (3 * 10 ^ 6))
1120 YS = 96 - Y * (45 / (3 * 10 ^ 6))
1130 PSET(XS,YS)
1140 X = X1:Y = Y1
1150  NEXT T
2000  END
```

D.14 Program (5.5)

```
90  REM  PROGRAM (5.5)
100  REM   ***** SET UP GRAPHICS CHARACTERISTICS *****

110   SCREEN 2:CLS
300   REM
                     *****  SET UP SCREEN DISPLAY *****

310   LINE(320,1)-(320,190)
320   FOR X = - 9 TO 9
330 XS = 320 + X * 101 / 3: REM    SCALE SCREEN HORIZONTALLY(30
PIXELS=1*10^6M)
```

D.14 PROGRAM 5.5

```basic
340  LINE(XS,96 + 3)-(XS,96 - 3): LOCATE 13,XS/8: PRINT X
350  NEXT X
360  LOCATE 14,44: PRINT "DISTANCE (10^6m)"
500  REM                       ***** SPECIFY INITIAL CONDITIONS *****
510  MM = 7.4 * 10 ^ 22: REM    MASS OF THE MOON (KG)
520  M = 7.4 * 10 ^ 22: REM    MASS OF SATELLITE (KG)
530  G = 6.67 * 10 ^ - 11: REM    UNIVERSAL GRAVITATIONAL CONSTANT
540  VA = 930: REM  VELOCITY OF MOON (HORIZONTAL MOTION RELATIVE TO THE SCREEN)
545  VB = 0
550  VX = 0: REM    INITIAL HORIZONTAL COMPONENT OF VELOCITY OF SATELLITE (RELATIVE TO SCREEN)
560  VY = 0: REM      INITIAL VERTICAL COMPONENT OF VELOCITY OF SATELLITE (RELATIVE TO SCREEN)
570  A = - 9 * 10 ^ 6: REM      INITIAL X COORDINATE OF MOON
580  B = 0: REM      INITIAL Y COORDINATE OF MOON
590  X = 0: REM    INITIAL X COORDINATE OF SATELLITE
600  Y = - 3 * 10 ^ 6: REM    INITIAL Y COORDINATE OF SATELLITE
610  DT = 50: REM          TIME INCREMENT
1000 REM                       ***** CALCULATE AND PLOT VELOCITY AND POSITION *****
1010  FOR T = 0 TO 15000 STEP DT
1030  R = SQR ((A - X) ^ 2 + (B - Y) ^ 2): REM    DISTANCE BETWEEN MOON AND SATELLITE (M)
1040  F = G * MM * M / (R * R): REM    GRAVITATIONAL FORCE BETWEEN MOON AND SATELLITE
1050  FX = F * (A - X) / R: REM     X-COMPONENT OF GRAVITATIONAL FORCE
1060  FY = F * (B - Y) / R: REM     Y-COMPONENT OF GRAVITATIONAL FORCE
1070  VX = VX + (FX / M) * DT:VA = VA - (FX / MM) * DT
1080  VY = VY + (FY / M) * DT:VB = VB - (FY / MM) * DT
1090  X1 = X + VX * DT:A1 = A + VA * DT
1100  Y1 = Y + VY * DT:B1 = B + VB * DT
1110  XS = 320 + X * (101 / (3 * 10 ^ 6)):AS = 320 + A * (101 / (3 * 10 ^ 6))
1120  YS = 96 - Y * (45 / (3 * 10 ^ 6)):BS = 96 - B * (45 / (3 * 10 ^ 6))
1130  PSET(XS,YS): PSET(AS,BS)
1140  X = X1:Y = Y1:A = A1:B = B1
1150  NEXT T
2000  END
```

D.15 Program (5.6)

```
90   REM   PROGRAM (5.6)
100  REM     ***** SET UP GRAPHICS CHARACTERISTICS *****

110  SCREEN 2:CLS
300  REM
                     *****  SET UP SCREEN DISPLAY *****

310  LINE(380,20)-(430,20): LOCATE 2,54: PRINT ")  (+X direction)":
LOCATE 2,45: PRINT "E"
320  LOCATE 4,45: PRINT "B": LOCATE 4,48: PRINT ".": LOCATE 4,56:
PRINT "(+Z direction)"
330  FOR A = - 2 TO 2
335  LINE(1,96)-(630,96)
340  X = 320 + A * 112
350  LINE(X,96 + 3)-(X,96 - 3):LOCATE 13,X/8: PRINT A
360  NEXT A
370  LINE(320,8)-(320,190): LOCATE 1,40: PRINT "Y"
380  LOCATE 14,50: PRINT "X(10^-4m)"
490
                     ***** SET UP INITIAL CONDITIONS *****

500  Q = 1.6 * (10 ^ - 19)
510  M = 9.1 * (10 ^ - 31)
520  E = 300
530  B = .01
540  V1 = 70000
550  V3 = 70000
560  DT = 1 * (10 ^ - 10)
570  X = 0:Y = 0
1000 REM
                 ***** CALCULATE AND PLOT VELOCITY AND POSITION *****

1010 V1 = V1 + (Q / M) * (E - V3 * B) * DT / 2
1020 V3 = V3 + (Q / M) * (V1 * B) * DT / 2
1030  FOR T = 0 TO 1.1 * 10 ^ - 8 STEP DT
1040 X1 = X + V1 * DT
1050 Y1 = Y + V3 * DT
1060 V2 = V1 + (Q / M) * ((E) - (V3 * B)) * DT
1070 V4 = V3 + (Q / M) * (V2 * B) * DT
1080 XS = 320 + X* 1.12 * (10 ^ 6)
1090 YS = 180 - Y * (.5 * 10 ^ 6)
1100  PSET(XS,YS)
1110 X = X1:Y = Y1:V1 = V2:V3 = V4
```

```
1120    NEXT T
2000    END
```

D.16 Program (6.1)

```
9   REM     PROGRAM (6.1) PROJECTILE IN ROTATING SYSTEM WITH UNIFORM
ANGULAR ACCELERATION
10  REM     THIS PROGRAM USES EULER'S ANGLES FOR ROTATION
80  REM
                ***** SET UP GRAPHICS CHARACTERISTICS *****


90  SCREEN 2:CLS
95  REM
                ***** SET UP SCREEN DISPLAY *****

100 REM         ANGLE A1 IS PHI:REM ANGLE A2 IS THETA:REM ANGLE A3 IS
PSI:REM ALL ANGLES ARE EXPRESSED IN RADIANS
110 A1 = 1.57
120 A2 = 1.57
130 A3 = 0
135   LOCATE 2,70:  PRINT "A1=";A1
137   LOCATE 3,70:  PRINT "A2=";A2
139   LOCATE 4,79:  PRINT "A3=";A3
140 S1 =   SIN (A1):C1 =   COS (A1)
150 S2 =   SIN (A2):C2 =   COS (A2)
160 S3 =   SIN (A3):C3 =   COS (A3)
170   REM                                                   L1  L2  L3
180   REM  CALCULATE ELEMENTS OF TRANSFORMATION MATRIX      L4  L5  L6
190   REM                                                   L7  L8  L9
200 L1 = C3 * C1 - C2 * S1 * S3
210 L2 =  - S3 * C1 - C2 * S1 * C3
220 L3 = S2 * S1
230 L4 = C3 * S1 + C2 * C1 * S3
240 L5 =  - S3 * S1 + C2 * C1 * C3
250 L6 =  - S2 * C1
260 L7 = S3 * S2
270 L8 = C3 * S2
280 L9 = C2
290 D = 20: REM          DISTANCE OF VIEWER FROM SCREEN (CM)
300 XV = D * L1
310 YV = D * L2
320 ZV = D * L3
330   READ A,X1,Y1,Z1,X,Y,Z
340   IF A = 16 GOTO 500
```

```
350   GOSUB 2000
390 SX = XS:SZ = ZS:X = X1:Y = Y1:Z = Z1
395   GOSUB 2000
400 LINE(XS,ZS)-(SX,SZ)
410   IF A = 4 THEN  LOCATE ZS/8,XS/8: PRINT "X"
420   IF A = 11 THEN  LOCATE ZS/8,XS/8: PRINT "Y"
430   IF A = 15 THEN LOCATE ZS/8,XS/8: PRINT "Z"
440   GOTO 330
500   REM
             ***** SET UP INITIAL CONDITIONS *****

505 YW = .1
510 W = 0
520 G = 0
530 X5 = 300:Y5 = 300:Z5 = 0
540 V1 = 0:V2 = 0:V3 = 0
550 DT = .01
1000  REM
             *****   CALCULATE AND PLOT VELOCITY AND POSITION   *****

1010  FOR T = 0 TO 23 STEP DT
1015 W = YW * T
1020 V4 = V1 + (W * W * X5 + 2 * V2 * W + Y5 * YW) * DT
1030 V5 = V2 + (W * W * Y5 - 2 * V4 * W - X5 * YW) * DT
1040 V6 = V3 - G * DT
1050 X6 = X5 + V1 * DT
1060 Y6 = Y5 + V2 * DT
1070 Z6 = Z5 + V3 * DT
1080 X = X5 / 100:Y = Y5 / 100:Z = Z5 / 100
1090  GOSUB 2000
1095  PSET(XS,ZS)
1100 V1 = V4:V2 = V5:V3 = V6
1110 X5 = X6:Y5 = Y6:Z5 = Z6
1120  NEXT T
1900  REM
                    ***** ROTATION SUBROUTINE *****


2000 XO = X - XV: REM      TRANSLATION OF COORDINATES TO MOVE
OBSERVER TO ORIGIN OF COORDINATES
2010 YO = Y - YV
2020 ZO = Z - ZV
2025  REM      ***** APPLY ROTATION MATRIX  ****
2030 X3 = L1 * XO + L2 * YO + L3 * ZO
2040 Y3 = L4 * XO + L5 * YO + L6 * ZO
```

```
2050 Z3 = L7 * X0 + L8 * Y0 + L9 * Z0
2060  REM     ***** PROJECT ROTATED OBJECT ON SCREEN *****
2070 XS = 320 + 45 * D * Y3 / ( - X3)
2080 ZS = 96 - 20 * D * Z3 / ( - X3)
2090  RETURN
3000  REM
                    ***** DATA FOR SCREEN DISPLAY *****
3010  DATA  1,3,3,0,-3,3,0
3020  DATA  2,3,2,0,-3,2,0
3030  DATA  3,3,1,0,-3,1,0
3040  DATA  4,4,0,0,-4,0,0
3050  DATA  5,3,-1,0,-3,-1,0
3060  DATA  6,3,-2,0,-3,-2,0
3070  DATA  7,3,-3,0,-3,-3,0
3080  DATA  8,3,3,0,3,-3,0
3090  DATA  9,2,3,0,2,-3,0
3100  DATA  10,1,3,0,1,-3,0
3110  DATA  11,0,4,0,0,-4,0
3120  DATA  12,-1,3,0,-1,-3,0
3130  DATA  13,-2,3,0,-2,-3,0
3140  DATA  14,-3,3,0,-3,-3,0
3150  DATA  15,0,0,4,0,0,0
3160  DATA  16,0,0,0,0,0,0
5000  END
```

D.17 Program (6.2)

```
5   REM PROGRAM (6.2)
10  REM   THIS PROGRAM USES EULER'S ANGLES FOR ROTATION
80  REM
                ***** SET UP GRAPHICS CHARACTERISTICS *****


90  SCREEN 2:CLS
95  REM
                ***** SET UP SCREEN DISPLAY *****

100 REM       ANGLE A1 IS PHI:REM ANGLE A2 IS THETA:REM ANGLE A3 IS
PSI:REM ALL ANGLES ARE EXPRESSED IN RADIANS
110 A1 = 0
120 A2 = 0
130 A3 = 0
135   LOCATE 2,70: PRINT "A1=";A1
137   LOCATE 3,79: PRINT "A2=";A2
139   LOCATE 4,70: PRINT "A3=";A3
```

APPENDIX D

```
140 S1 = SIN (A1):C1 = COS (A1)
150 S2 = SIN (A2):C2 = COS (A2)
160 S3 = SIN (A3):C3 = COS (A3)
170 REM                                                    L1  L2  L3
180 REM   CALCULATE ELEMENTS OF TRANSFORMATION MATRIX      L4  L5  L6
190 REM                                                    L7  L8  L9
200 L1 = C3 * C1 - C2 * S1 * S3
210 L2 =  - S3 * C1 - C2 * S1 * C3
220 L3 = S2 * S1
230 L4 = C3 * S1 + C2 * C1 * S3
240 L5 =  - S3 * S1 + C2 * C1 * C3
250 L6 =  - S2 * C1
260 L7 = S3 * S2
270 L8 = C3 * S2
280 L9 = C2
290 D = 20: REM        DISTANCE OF VIEWER FROM SCREEN (CM)
300 XV = D * L1
310 YV = D * L2
320 ZV = D * L3
330  READ A,X1,Y1,Z1,X,Y,Z
340  IF A = 16 GOTO 500
350  GOSUB 2000
390 SX = XS:SZ = ZS:X = X1:Y = Y1:Z = Z1
395  GOSUB 2000
400  LINE(XS,ZS)-(SX,SZ)
410  IF A = 4 THEN LOCATE ZS/8,XS/8: PRINT "X"
420  IF A = 11 THEN LOCATE ZS/8,XS/8: PRINT "Y"
430  IF A = 15 THEN LOCATE ZS/8,XS/8: PRINT "Z"
440  GOTO 330
500  REM                ***** SET UP INITIAL CONDITIONS *****

505 X0 = SX:Z0 = SZ
510 SI = 0:VS = 100
520 PH = 0:VP = 6: REM     PH IS PHI AND VP IS ANGULAR VELOCITY
ASSOCATED WITH PHI
530 TH = .6:VT = 6
540 I3 = 1000: REM  MOMENT OF INERTIA
550 I2 = 3000:I1 = 3000
560 M = 100
570 H = 5
580 G = 980
590 PP = ((I2 * ( SIN (TH) ^ 2) + I3 * ( COS (TH) ^ 2)) * VP) + I3
 *  COS (TH) * VS
600 PS = I3 * (VS + VP *  COS (TH))
```

```
610 DT = .005
1000  REM
                ***** CALCULATE AND PLOT VELOCITY AND POSITION *****

1010  FOR T = 0 TO 10 STEP DT
1020  VS = (PS / I3) - ((PP - PS *  COS (TH)) *  COS (TH)) / (I1 * ( SIN (TH) ^ 2))
1030  VP = ((PP) - (PS *  COS (TH))) / (I1 * ( SIN (TH) ^ 2))
1040  V1 = VT + ((((I1 - I3) * (VP ^ 2) *  COS (TH) - I3 * VS * VP + M * G * H) *  SIN (TH)) / I1) * DT
1060  P1 = PH + VP * DT
1070  T1 = TH + V1 * DT: REM    LAST POINT APPROXIMATION
1072  SI = S1:PH = P1:TH = T1:VT = V1
1080  X = 3 * ( SIN (TH) *  COS (PH))
1090  Y = 3 * ( SIN (TH) *  SIN (PH))
1100  Z = 3 *  COS (TH)
1110   GOSUB 2000
1120   PSET(XS,ZS)
1140   NEXT T
1900   REM
                    ***** ROTATION SUBROUTINE *****

2000  XO = X - XV: REM      TRANSLATION OF COORDINATES TO MOVE
OBSERVER TO ORIGIN OF COORDINATES
2010  YO = Y - YV
2020  ZO = Z - ZV
2025  REM      ***** APPLY ROTATION MATRIX ****
2030  X3 = L1 * XO + L2 * YO + L3 * ZO
2040  Y3 = L4 * XO + L5 * YO + L6 * ZO
2050  Z3 = L7 * XO + L8 * YO + L9 * ZO
2060  REM     ***** PROJECT ROTATED OBJECT ON SCREEN *****
2070  XS = 320 + 45 * D * Y3 / ( - X3)
2080  ZS = 96 - 20 * D * Z3 / ( - X3)
2090   RETURN
3000   REM
                   ***** DATA FOR SCREEN DISPLAY *****

3010   DATA   1,3,3,0,-3,3,0
3020   DATA   2,3,2,0,-3,2,0
3030   DATA   3,3,1,0,-3,1,0
3040   DATA   4,4,0,0,-4,0,0
3050   DATA   5,3,-1,0,-3,-1,0
3060   DATA   6,3,-2,0,-3,-2,0
3070   DATA   7,3,-3,0,-3,-3,0
```

```
3080   DATA    8,3,3,0,3,-3,0
3090   DATA    9,2,3,0,2,-3,0
3100   DATA    10,1,3,0,1,-3,0
3110   DATA    11,0,4,0,0,-4,0
3120   DATA    12,-1,3,0,-1,-3,0
3130   DATA    13,-2,3,0,-2,-3,0
3140   DATA    14,-3,3,0,-3,-3,0
3150   DATA    15,0,0,4,0,0,0
3160   DATA    16,0,0,0,0,0,0
5000   END
```